Alexandra Carina Gruber
Biotech Funding Trends

Related Titles

Behme, S.

Manufacturing of Pharmaceutical Proteins

2009
ISBN: 978-3-527-32444-6

Borbye, L.

Industry Immersion Learning

Biotechnology Cases for Graduate Students in Science and Business

2009
ISBN: 978-3-527-32408-8

Wink, M. (ed.)

An Introduction to Molecular Biotechnology

Molecular Fundamentals, Methods and Applications in Modern Biotechnology

2006
ISBN: 978-3-527-31412-6

Alexandra Carina Gruber

Biotech Funding Trends

Insights from Entrepreneurs and Investors

WILEY-VCH Verlag GmbH & Co. KGaA

The Author

Dr. Alexandra Carina Gruber
Vienna, Austria

Library of Congress Card No.:
applied for

British Library Cataloguing-in-Publication Data
A catalogue record for this book is available from the British Library.

Bibliographic information published by the Deutsche Nationalbibliothek
The Deutsche Nationalbibliothek lists this publication in the Deutsche Nationalbibliografie; detailed bibliographic data are available on the Internet at http://dnb.d-nb.de.

© 2009 WILEY-VCH Verlag GmbH & Co. KGaA, Weinheim

Editing   Gabriele Berghammer, the text clinic, Vienna
Typesetting   Thomson Digital, Noida, India
Printing   betz-druck GmbH, Darmstadt
Binding   Litges & Dopf Buchbinderei GmbH, Heppenheim
Cover Design   Adam-Design, Weinheim
Alexander Schatek, Wr. Neustadt

Printed in the Federal Republic of Germany
Printed on acid-free paper

ISBN: 978-3-527-32435-4

Cover
Cover picture with kind permission of *onepharm Research & Development GmbH*, Vienna.

For my parents
who were the most important teachers
during much of my own development

"It is not the strongest of the species that survive,
nor the most intelligent,
but the ones most responsive to change." (Charles Darwin)

Contents

Biotech Funding Trends: Insights from Entrepreneurs and Investors.
Alexandra Carina Gruber
Copyright © 2009 WILEY-VCH Verlag GmbH & Co. KGaA, Weinheim
ISBN: 978-3-527-32435-4

Preface

The European biotech industry has long lagged behind its US counterpart. For most of Europe, the same holds true for private equity and venture capital (VC) spending on innovative start-ups. In recent years, however, the European biotech industry has improved its competitive position vis-à-vis the US, and so has the development of the private equity and VC market. Despite this positive trend, only a small fraction of available capital is invested in biotech start-ups – probably because there is hardly an area more complex, more globally oriented, more time-consuming, and more risky than biotech. As a result, many biotech start-ups are still struggling for survival, particularly in cultures that are less entrepreneurially and private-equity driven.

Has venture capital indeed developed into the dominant financing form for biotech, as it had for other start-up industries, such as information technology? How do above trends affect the interaction between European entrepreneurs, venture capitalists, and other investors? Is the value of venture capital perceived differently between entrepreneurs and investors? Are there any other important financial tools for biotech start-ups? What are the most important advantages and shortcomings of the different financing forms available? What makes the ideal entrepreneur or investor? Why do so many biotech start-ups fail? Can improved communication between entrepreneurs and their investors as well as among investors themselves help foster mutual understanding and lead to more successful long-term partnerships?

Answers to these and other questions about the interpersonal and financial aspects of starting a biotech company were obtained by conducting qualitative interviews with three groups of interview partners, i.e., entrepreneurs, venture capitalists, and other investors. The results of this prospective research study are reported in this book. Entrepreneurs and venture capitalists both inhabit two complex and differing ecosystems with many important, but often opaque, rules and interactions. In view of their different backgrounds, entrepreneurs and investors need to first find a way to understand – and begin to use – the language spoken by their respective counterparts. Interests between different groups of investors, too, can differ substantially. It is therefore nothing short of a challenge to reconcile the often diverging interests of all key players involved in starting a biotech company.

Biotech Funding Trends: Insights from Entrepreneurs and Investors.
Alexandra Carina Gruber
Copyright © 2009 WILEY-VCH Verlag GmbH & Co. KGaA, Weinheim
ISBN: 978-3-527-32435-4

By describing the characteristics, beliefs, and expectations of each of these groups, this book intends to support both entrepreneurs and investors in getting along more successfully with each other. At the end of the day, finding the most suitable partner and building a relationship based on mutual understanding greatly increases the chances of a biotech start-up to succeed in a fiercely fought-over market.

N.B.: For easier reading, only male titles were used.

Acknowledgments

I would like to take the opportunity to thank a lot of people who have, directly or indirectly, been involved in writing this book.

First of all, I would like to extend a warm thank you to Professor Thomas Hellmann from the University of British Columbia (UBC), Vancouver (www.strategy.sauder.ubc.ca/hellmann/). It was more than an exciting experience to work with Thomas on the development of my master thesis, both conceptually and practically. Thomas gave me lots of insights into the world of global venture capital. First inspired by the issue during Thomas' lectures in November 2006, the idea of writing a thesis – and later this book – on venture capitalism has now come to fruition.

Second, I would like to thank all of my interview partners, without whom this book would not have been possible. It was exciting and challenging to interview so many highly experienced entrepreneurs, venture capitalists, and other experts and investors, and discuss their insights into the European biotech market and the financing models available to develop a company. Every single interview contributed distinctively to the overall conclusions of this work, which would have been unthinkable without their generous and valuable support. Special thanks go to the contributors of the extended biotech case studies presented in this book, namely to Thomas Fischer and Susanne Bach of AUSTRIANOVA Biomanufacturing AG, Michael Tscheppe, Reinhard Zickler, and Isolde Bergmann of AVIR Green Hills Biotechnology AG, Wolfgang Schönfeld of EUCODIS GmbH, Werner Lanthaler of Intercell AG, and Bernhard Küenburg of onepharm GmbH.

Last but not least, I wish to thank my parents, who have given me unlimited support both during my educational and professional career and while writing this book. I feel privileged to have been able to receive such high-level education, and I believe this journey is bound to continue.

Moreover, representative of the many friends and colleagues who supported me, let me mention my friend and English editor Gabi (www.the-text-clinic.com), my graphic artist Alexander (www.schatek.at), my friend and former classmate Teodoro, and my good old friend Herbert. It is an incredible gift to be able to rely on the knowledge of such highly experienced and talented friends. While working

Biotech Funding Trends: Insights from Entrepreneurs and Investors.
Alexandra Carina Gruber
Copyright © 2009 WILEY-VCH Verlag GmbH & Co. KGaA, Weinheim
ISBN: 978-3-527-32435-4

on this book, I was able to follow their recommendations in so many different ways.

Therefore, dear reader, please consider this piece of work a heartfelt thank you to my instructor, to my many interview partners, and to my family and friends. Thank you all – and please enjoy reading!

Vienna, July 2008   *Alexandra Carina Gruber*

Limit of Liability/Disclaimer of Warranty

While the publisher and author have used their best efforts in preparing this book, they make no representations or warranties with respect to the accuracy or completeness of the content of this book and specifically disclaim any implied warranties of merchantability or fitness for a particular purpose. No warranty may be created or extended by sales representatives or written sales materials. Rather, the assumptions expressed in this book reflect the personal interpretation of the authors and are subject to the continuous changes of research, (clinical) experience, and scientific opinion. The advice and strategies contained herein may not be suitable for the reader's situation, who should consult with a professional, where appropriate. Neither the publisher nor author shall be liable for any loss of profit or any other commercial damages, including but not limited to special, incidental, consequential, or other damages.

Designations used by companies to distinguish their products are often claimed as trademarks. In all instances where John Wiley & Sons, Inc., is aware of a claim, the product name appears with initial capital or all capital letters. Readers, however, should contact the appropriate companies for more complete information regarding trademarks and registration. The absence of trademark or registration symbols should not be taken to indicate absence of trademark protection; anyone wishing to use product names in the public domain should first clear such use with the product owner.

All websites mentioned as references were accessed between January and July 2008.

Biotech Funding Trends: Insights from Entrepreneurs and Investors.
Alexandra Carina Gruber
Copyright © 2009 WILEY-VCH Verlag GmbH & Co. KGaA, Weinheim
ISBN: 978-3-527-32435-4

Executive Summary

The European biotech industry has long lagged behind its US counterpart. For most of Europe, the same was true for private equity and venture capital spending on innovative start-ups. In recent years, however, the European biotech industry has improved its competitive position vis-à-vis the US, and so has the development of the private equity and venture capital market. Despite this positive trend, only a small fraction of available capital is invested in biotech – probably because there is hardly an area more complex, more globally oriented, more time-consuming, and more risky. As a result, many biotech start-ups today are still struggling for survival, particularly in cultures that are less entrepreneurially and private-equity driven, such as Austria and Germany, where there is still an evident gap between seed and early-stage financing. Considering the long lead times involved in biotechnological research and development, the unwillingness of venture capitalists to fund early development may largely be due to start-ups being unable to propose viable exits. At the same time, it has become tougher for companies to raise venture capital, as consolidation has thinned the ranks of venture capital firms.

This book focuses on two counteracting trends: One is the private equity and venture capital spending boom and the increasing number of biotech start-ups with strong product pipelines. At the same time, compared to other industries, the amount of venture capital spent on biotech start-ups is still limited.

In general, each development phase of a young company is financed by specific types of investors. In the preseed and seed phases, financing in many cases starts with the 'three Fs' – family, friends, and fools. After that, it is usually the government that steps in with grants and loans, followed by business angels and early-stage venture capitalists. In later development stages and as capital requirements increase, strategic alliances with corporate companies are key to success, as are deals with late-stage venture capitalists or banks. The last link in the financing chain of a biotech start-up is its exit, usually in form of an initial public offering (IPO), trade sale, or management buy-out.

Based on the general financing matrix for start-up companies and given the complexity of biotech development, the primary question of this survey was whether a biotech start-up went through the same development phases and was financed by

Biotech Funding Trends: Insights from Entrepreneurs and Investors.
Alexandra Carina Gruber
Copyright © 2009 WILEY-VCH Verlag GmbH & Co. KGaA, Weinheim
ISBN: 978-3-527-32435-4

the same types of investors as other innovative start-up companies, such as IT or industrial product companies. Do all biotech companies share the same financing model? What do investors think about the relative importance of available financing options? Which financing models are used in early-stage start-ups, and which are used in later stages? What do biotech entrepreneurs and investors consider to be the major advantages and drawbacks of available financing sources? How can the gap between seed and follow-on financing be closed? Are there any new and creative financing approaches available to young biotech companies? What do entrepreneurs and investors think about the different exit scenarios available?

Another focus of this project was to determine what the interpersonal relationships and alliances in the field of biotechnology looked like, what the characteristics of the individual key players were, and how they interacted with one another. How can early- and late-stage alliances be set up most successfully and what is expected of an investor to best support a young biotech company? How can networking between the individual players be intensified and how significant are such networks? Why have many biotechnology start-ups failed?

To answer these questions, standardized qualitative interviews were carried out with three groups of interview partners, i.e., entrepreneurs, venture capitalists, and investors other than venture capitalists. The third group of other investors was considered a control group of sorts, because they are usually good observers of the relationship between the entrepreneur and venture capitalist. Analysis of the available funding and networking options and their relative importance from the perspectives of the major players in this industry finally enabled practical guidelines for new entrepreneurs and investors to be developed, giving insight into the lessons learned by their peers who had gone before them and helping improve mutual understanding between the parties involved in the successful launch of a biotechnology company.

In terms of the strengths and abilities required of an entrepreneur, views across groups were surprisingly homogeneous, considering that entrepreneurs and investors come from two different worlds – science and business – potentially making communication between the two something of a challenge. The most important attributes of an entrepreneur were leadership and interpersonal skills, followed by experience, knowledge, and conviction. When starting a company, it will be beneficial for the entrepreneur to acquire management skills to better understand the mechanisms behind a business plan and the interplay of the various forces driving an enterprise, such as market, competition, and pricing. From the very outset, the vision of the entrepreneur should be to establish a business which fulfils an underserved market need and which is global and sustainable. Later on, as the company structure becomes more diversified, complex, and demanding, additional staff with complementary skills will have to come on board to lead the company into the next development phase. Different development stages call for different skill sets to have the company value increase in the most effective and sustainable manner. Failure of any member of the management team to understand this may at times require tough personnel decisions to be taken and management to be reshuffled. Delaying such decisions may put the success of the entire company on the line.

Regarding the major strengths and abilities required of a venture capitalist, a dense network of relations, experience, and knowledge ranked first in all three groups of interviewees. In addition, entrepreneurs considered it important to partner with a venture capitalist who has both enough capital to spend and a long-term perspective. Other essential factors were the size of available funds, access to financial markets, know-how, and experience both in the pharmaceutical industry and in deal making.

The analysis of the most important financing forms also produced homogenous results across groups, with the major sources of biotech funding being venture capital, government grants, and strategic alliances. Venture capital firms are highly target- and return-oriented, pushing the limits of a company. With venture capital often referred to as 'smart money,' these companies also offer added value in matters of management and business administration. In terms of drawbacks, venture capitalists themselves confirmed that funds were hard to come by, with fierce competition among applicants and only a small fraction of start-ups actually receiving funding. Equity participation dilutes the entrepreneur, who may as a result have to abandon certain rights. It is therefore essential for the founder to be given adequate incentives and sufficient shares to motivate him to stay in the company.

With regard to government funding, results were rather heterogeneous between the groups interviewed and the countries where their companies were located. Generally, entrepreneurs and venture capitalists considered this source of funding less important than did the group of other investors. Government grants are generally considered money that is easy to obtain, inexpensive, and non-dilutive for the entrepreneur. Government funds are usually entrepreneur-friendly and unbureaucratic. They dispose over significant fund sizes, usually without applying much due diligence. On the downside, government funds may wish to bring their influence to bear on the company's management team. While providing significant start-up capital, they sometimes lack the business know-how required to efficiently support a young entrepreneur. When debt financing is involved, the entrepreneur may find himself confronted with the challenge of having to pay back a loan at a time when additional, originally unexpected financial challenges hit the company.

Strategic alliances enable intensive scientific cooperation, with the added value of validating the newly developed technology or product. Alliances provide non-dilutive capital and access to existing marketing and distribution channels and are therefore a fantastic instrument for commercialization of the new product. It is a sign of quality for a young start-up to be taken seriously by one of the major pharmaceutical players, and this may help attract new investors. On the downside, partnering with a big player too early may be tantamount to giving away much of the company's assets at a significant discount. Well established pharmaceutical companies, having to follow their internal 'chain of command,' may be slower to react than other types of investors. Also, pharmaceutical companies may be less interested in promoting the sustainable growth and wellbeing of the start-up than in securing their own share, e.g., if a newly developed and promising molecule is at stake. As a result, the entrepreneur may have to abandon much – or all – of his freedom and independence.

The picture for additional financing sources was more heterogeneous across groups. Entrepreneurs are very innovative when it comes to financing expensive research and development projects, and almost every entrepreneur interviewed appeared to have taken an individual approach to financing their respective companies. One example that was frequently mentioned by entrepreneurs from German-speaking countries is a debt-financing form called 'atypical silent partnership.' For venture capitalists, business angel financing was an important alternative, whereas the group of other experts regarded EU grants as growing in significance.

The company valuation preferences likewise were rather diverse. Thus, in earlier phases, the valuation techniques considered acceptable by the interviewees are manifold, ranging from comparables, the venture capital method, valuation of the management, and gut feeling to discounted cash flow (DCF), real options, and even Monte Carlo simulations. In later development stages, there is a shift to fewer and more sophisticated techniques, i.e., DCF, Monte Carlo analysis, or real options. Overall, a universally applicable valuation method does not exist. Therefore, applying various techniques will result in a more differentiated picture of the overall company value, and combining various methods was described as the most meaningful option by the interview partners.

Early-stage partnerships between investors and universities or their scientists appear to be of particular interest to entrepreneurs planning a university or company spin-off, increasing the likelihood of getting access to early funding, such as government grants and venture capital. With regard to business angels as early-stage financing option, most entrepreneurs and venture capitalists believe that their importance will increase. However, some venture capitalists also see some potential for conflict, with business angels perhaps wanting to maintain their influence in the company in subsequent, bigger financing rounds without bringing in additional capital.

Late-stage alliances between venture capitalists and independent funds made available by major pharmaceutical companies, such as the SR-One Fund by GlaxoSmithKline or the Novartis Venture Funds, are generally viewed positively, as long as the pharmaceutical company feeding the fund does not bring any particular strategic influences to bear on the deal, e.g., by claiming special rights. Such rights could lead to a conflict of interest. Thus, on the one hand, a corporate venture capital fund is accountable to his investor, i.e., the mother company. On the other, he is obliged to sell the company at the best achievable market price. All three groups agreed on the role of banks in financing start-ups. Because biotech start-ups do not offer banks a security guarantee and banks themselves lack the scientific and technological expertise necessary to meaningfully support biotech research, the risk to the bank of investing into a biotech start-up is incalculable. The importance of banks as capital investors changes in the phase preceding an IPO, which banks sometimes support with mezzanine capital.

Most entrepreneurs interviewed did not have a clear exit vision. However, their general goal was an exit strategy guaranteeing the highest achievable price combined with the highest level of personal independence. For the venture capitalist, the most preferred exit route was the trade sale.

All interview partners openly talked about the potential drawbacks of the financing forms they themselves offered. For example, venture capitalists were well aware that, by bringing in money and thereby 'diluting' the entrepreneur's stakes in the company, his motivation was likely to decrease. All three groups agreed on the major biotech financing forms and their relative strengths and weaknesses. In contrast, the questionnaire items dealing with alliances and networking opportunities produced rather heterogeneous results between the different groups of investors. Therefore, in addition to having to identify an entrepreneur who best fits his investment philosophy, an additional challenge for the venture capitalist is to team up with suitable follow-on or co-investors to increase the likelihood of a successful exit.

In the past, many entrepreneurs struggled to obtain venture capital funding. Fortunately, the biotech scene has changed and venture capital is on the rise. Yet, its importance is increasing at a slower than expected rate, and the financing gap between early government funding and follow-on venture capital is still wide open. Encouragingly, the European biotech scene is bringing forth more and more serial entrepreneurs, capable of building upon their earlier successes in leading companies to an IPO or trade sale. This knowledge, coupled with the experience gained through networking among the different groups of investors, will lay the groundwork necessary for having a fledgling company develop into one with a convincing product portfolio. The pipelines of biotech start-ups are definitely there waiting to be exploited by dedicated entrepreneurs and experienced venture capitalists. With trust in the biotech industry and among its key players strengthening, the opportunities for start-ups to develop into sustainable businesses have improved and should be well poised to bring innovative new products to global markets.

1
Introduction

1.1
Development of the Private Equity and Venture Capital Market

1.1.1
Definition

Private equity is an asset class consisting of many forms of high-risk and high-return equity investments into companies not traded on a public stock exchange. Private equity deals are generally thought of as a long-term investment (International Financial Services London, 2006). Investments typically involve a transformational, value-added, active management strategy helping start-ups improving their fundamental business practices. By that means, private equity may be considered as one of the most potent forces enhancing economy-wide improvements in corporate productivity. Private equity firms pursue one major goal – to search for companies with the potential for growth and to put in place the capital, talent, and strategy required to permanently strengthen the company and raise its value. Private equity is often subsumed under the umbrella of 'alternative investments,' complementing the stock and bond portfolios traditionally used by investors (European Venture Capital Association, 2007a). Private equity can further be subdivided into venture capital, mezzanine capital, buy-outs, and turnarounds/distress (Metrick, 2007).

Venture capital, often also referred to as 'smart money' because it comes in a combination of know-how and a network of experts, by definition refers to private equity investments made for the launch, early development, or expansion of a business. Venture capital focuses on new entrepreneurial undertakings rather than on mature businesses (European Venture Capital Association, 2007a). It represents only a small proportion of the overall private equity market. Venture capitalists act as financial intermediaries, who take the investors' capital and invest it directly in portfolio companies (Metrick, 2007). Investments can be classified into seed-stage, start-up, expansion-stage, and replacement capital. Seed-stage capital is financing provided to research, assess, and develop an initial concept before a business has reached the start-up phase. Start-up stage capital is financing for product development and initial marketing, expansion-stage capital is financing for growth of a

Biotech Funding Trends: Insights from Entrepreneurs and Investors.
Alexandra Carina Gruber
Copyright © 2009 WILEY-VCH Verlag GmbH & Co. KGaA, Weinheim
ISBN: 978-3-527-32435-4

company which is close to break-even or trading profitably, and replacement capital is used to purchase shares from another investor or to reduce gearing via the refinancing of debt (International Financial Services London, 2006). A special form of venture capital is corporate venture capital, a term used to describe the investment of corporate funds directly in external start-up companies, which are then free to operate independently and to make autonomous investment decisions. These corporate venture capital firms will often be led by strategic objectives other than financial results and will have neither dedicated capital nor an expectation that capital will be returned within a certain time frame (Metrick, 2007). Examples of such corporate venture capital funds in biotech are the Novartis Venture Funds and the Glaxo-SmithKline SR–One Fund.

Mezzanine money has two different meanings in the private equity industry. The first refers to a form of late-stage venture capital, with financing typically occurring in the form of subordinated debt and some additional equity participation in the form of options to buy common stock. The second meaning of mezzanine first arose in the mid-1980s, when investors started to use the same capital structure, i.e., subordinated debt with some equity participation, to establish yet another layer of debt financing for highly leveraged buy-out transactions. Today, most private equity firms offering mezzanine capital practice the latter type of investing (Metrick, 2007).

Buy-outs, which occur when a private equity investment firm gains control of a majority of the company's equity through the use of debt, are by far the largest private equity segment. These are typically investments in more mature companies. The acquisition normally entails a change of ownership (International Financial Services London, 2006). In large buy-outs, the investors put up the equity stake, today usually between 20% and 40% of the total purchase price, and borrow the rest from public markets, banks, or mezzanine investors – hence the term leveraged buy-outs (LBO; Metrick, 2007).

Finally, turnarounds are investments into a distressed company or a company where value can be unlocked as a result of a one-time opportunity, such as changing industry trends or government regulations (Metrick, 2007).

Private equity funds typically have limited and general partners. The general partners are generally represented by the private equity investors who manage the fund. The limited partners are institutional and individual investors who provide capital. These are limited in the sense that their liability only extends to the capital that they contribute (Lerner and Gompers, 2001). The most common limited partners are institutional investors, such as pension funds, banks, insurance companies, or endowments. Each fund raised by a private equity company is invested in a number of firms with a five- to ten-year horizon. When a fund is ended, its cash proceeds, coming chiefly from initial public offerings (IPOs) and trade sales, are distributed to investors together with any remaining equity holdings. The number and variety of groups that invest in private equity have expanded substantially to include a wide range of different types of investors. Until two decades ago, the private equity market primarily consisted of wealthy individuals investing in early-stage companies. In recent years, the situation has changed and there are now many institutional investors with long-term commitments (International Financial Services London,

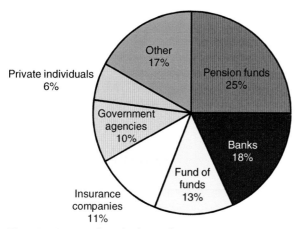

Figure 1.1 Sources of new funds raised in Europe (International Financial Services London, 2006).

2006). The distribution of the different private equity investor types in Europe is shown in Figure 1.1.

Private equity firms generally receive a return on their investments in one of three ways, i.e., an IPO, a trade sale or merger, or recapitalization (as in the form of a management buy-out). An IPO is the first sale of stock by a private company to the public. Such a stock market flotation may be the most spectacular exit, but it is far from being the most widely used, even in stock market booms. For an IPO to be successful, the company first has to invest enormous amounts of time and money into building a pipeline that is likely to attract investors. Furthermore, IPOs are associated not only with substantial one-time direct and indirect costs, but also with ongoing costs and the need to supply information on a regular basis to investors and regulators for publicly traded firms (Ritter, 1998). A stock market flotation should therefore reflect the genuine wish to make the company more dynamic over the long term and to profit from the growth possibilities offered by a stock market. Once the stock is publicly traded, this enhanced liquidity allows the company to raise capital on more favorable terms than if it had to compensate investors for the lack of liquidity associated with privately-held companies (European Venture Capital Association, 2007a). Also, existing shareholders are able to sell their shares in open-market transactions (Ritter, 1998). Overall, a flotation is not an end in itself, but the beginning of a long development process (European Venture Capital Association, 2007a).

Trade sales are a very popular alternative route for a divestment. A trade sale of a privately held company equity, also referred to as mergers and acquisitions (M&A), is the sale of company shares to industrial investors. Large and small companies often complement each other, and an alliance between them will not only serve to round off their portfolios, it will also guarantee a strategic advantage to at least one of the companies teaming up. This is why buyers are often willing to pay a premium to acquire a complementary business (European Venture Capital Association, 2007a). Although the terms 'merger' and 'acquisition' are often used as though they were synonymous, they do mean slightly different things. When one company takes over another and clearly establishes itself as the new owner, the purchase is called an acquisition. From a legal point of view, the target company ceases to exist, the buyer 'swallows' the business, and the buyer's stock continues to be traded. In the pure sense of the term, a merger happens when two firms, often of about the same size, agree to go forward as a single new company rather than remain separately owned and operated. Both companies' stocks are surrendered, and new company stock is issued in its place. In practice, however, actual mergers of equals do not happen too often. Usually, one company buys another and, as part of the deal's terms, simply allows the acquired firm to proclaim that the action is a merger of equals, even if it is technically an acquisition.

Recapitalizations are based on a company incurring significant additional debt by repurchasing stocks through a buyback program or by distributing a large dividend among its current shareholders. This causes the share price to soar, making the company a less attractive takeover target. Recapitalizations are mostly used to fend off a hostile acquisition. The repurchase of a company by its management team, the management buy-out (MBO), is becoming more and more successful as an exit strategy. It is a highly attractive option for both the investment manager and the company's management team, provided that the company can guarantee regular cash flows and mobilize sufficient loans (European Venture Capital Association, 2007a).

1.1.2
Historical Background

The seeds of the private equity industry were planted in 1946, when Harvard professor Georges Doriot created American Research and Development (ARD) together with Karl Compton, then president of the Massachusetts Institute of Technology, Merrill Griswold, former chairman of the Massachusetts Investors Trusts, and Ralph Flander, then president of the Federal Reserve Bank of Boston. ARD's overall aim was to raise funds from wealthy individuals and college endowments and to invest them in entrepreneurial start-ups in technology-based manufacturing. Its founders believed that by providing management with skills and funding, they could encourage companies to succeed – and in doing so, make a profit themselves (Bottazzi and Da Rin, 2002).

In the 1980s, FedEx and Apple were able to grow because of private equity and venture funding, as were Cisco, Genentech, Microsoft, Avis, Sun Microsystems, and many others (Bottazzi and Da Rin, 2002). Despite these successes, private equity companies came to be regarded with acrimony because of a series of debt-financed leveraged buy-outs (LBOs) of established firms, casting private equity firms as irresponsible corporate raiders and as a threat to the free capitalist structure (Burrough and Helyar, 2003).

LBOs were pioneered by Jerry Kohlberg, Henry Kravis, and George Roberts, who later founded the private equity firm Kohlberg Kravis Roberts (KKR). The idea was to employ aggressive forms of financial engineering to increase shareholder value. For this to work, certain management principles were to be followed, i.e., recapitalization of the company to substantially increase debt, a concentration on maximization of after-tax cash flow from operations, the sale of unnecessary assets, and motivation of the management with exceptionally high compensation incentives. After several years of intense management, they would take the improved company public again or sell it to a larger corporation, in this way returning capital and a double-digit return to investors (Smith, 2007).

In the first half of the 1980s, the annual flow of money into venture capital funds increased by a factor of ten, but it steadily declined from 1987 to 1991. Through the 1980s, the rise was even more dramatic for buy-outs, but this too was followed by a precipitous fall at the end of the decade, mostly due to changing fortunes of private equity investments. In the mid-1980s, returns on venture capital funds declined sharply. This fall was triggered by overinvestment in a few industries, such as computer hardware, and by many new and inexperienced venture capitalists entering the scene. A similar decline was seen for buy-out returns in the late 1980s, due largely to the increased competition between transactions. As investors became disappointed with returns, they spent less capital in the industry (Lerner et al., 2005).

The 1990s were characterized by similar patterns on an unprecedented scale. Much of the decade saw dramatic growth and excellent returns in the private equity industry. This recovery was due to the withdrawal of many inexperienced investors, ensuring that the remaining groups were facing less competition for transactions. Moreover, there was a healthy market for IPOs, guaranteeing easier exit for equity investors. Meanwhile, the extent of technological innovation created extraordinary opportunities for venture capitalists. New capital commitments to both venture and buy-out funds increased to record levels by the late 1990s and 2000. Yet, private equity grew at a pace that was too fast to be sustainable. Institutional and individual investors, attracted especially by the high returns of venture funds, flooded unprecedented amounts of money into the industry, often resulting in overchallenged partners, inadequate due diligence, and poor investment decisions (Lerner et al., 2005).

After reaching a peak in 2000, private equity funds fell between 2001 and 2004, mainly due to the slowdown in the global economy and declines in equity markets, particularly in the technology sector. In 2005, market confidence and trading conditions improved again, with US$135 billion of private equity invested globally, up one-fifth on the year 2004. Between 2000 and 2005, buy-outs generated a growing portion of private equity investments, increasing from one-fifth to more than two-thirds. By contrast, the share of venture capital investment fell during this period. Private equity fund raising also reached a peak value of US$232 billion, up three-quarters on 2004. In 2006, according to International Financial Services London, the positive trend continued, with a record amount of US$365 billion of private equity being invested globally and with private equity fund raising amounting to US$335 billion, both marking a significant upswing to the previous year (International Financial Services London, 2007).

Historically, Europe has had to grapple with more impediments to private equity than the US, not least because of tougher labor and bankruptcy laws. However, also in Europe, the ability of private equity investors to restructure, grow, and sell profitably has, over the past years, generated more and more interest from key players, such as debt providers, pension plans, and public company CEOs (Blaydon and Wainwright, 2007). Between 2000 and 2005, Europe increased its global share of private equity investments from 17% to 43% and raised funds from 17% to 38%. This was largely a result of strong buy-out market activity in Europe. In contrast, private equity activity in North America decreased from 68% to 40%, and raised funds dropped from 69% to 52% (International Financial Services London, 2006). This illustrates that private equity represents an increasingly important source of financing for European growth businesses. More than half of all private equity investments go into small and medium-sized companies with 100 staff or less, with these enterprises representing the main source of new jobs in many economies (Gaspar, 2007). On average, companies financed by private equity develop faster, invest more, and generate more jobs. Private equity firms themselves contribute a lot more than just funding, including hands-on management expertise, counseling, and access to regional and global networks. Without the involvement of private equity firms, many companies would not have been founded, reorganized, or found successive owners or managers (Frommann, 2007).

1.1.3
The Challenges to Private Equity in Europe

However, this does not mean that private equity is not facing challenges. Thus, fundraising in Europe is currently geared towards buy-outs, reducing the volume of available venture capital invested in early-stage development. More importantly, private equity is a boom-and-bust business that is highly sensitive to the whims of economic cycles. The past ten years have seen sharp fluctuations in the capital raised by private equity funds. The consequences of this are two-fold. First, capital availability to support private equity business growth can dry up rapidly in bear markets. Second and even more important, exit prices and the resulting returns to private equity investors are correlated with the global equity markets, decreasing the appeal of private equity as an investment instrument (Gaspar, 2007). Also, surveys have shown that the public generally knows very little about the methods and procedures of private equity and venture capital players. Therefore, if the capital base is to be improved, the industry will have to provide more information and improve its reputation by clearly informing the public of its true economic impact (Marchart, 2007).

1.1.4
Private Equity and Venture Capital Performance in Europe by Country

Despite these challenges, the overall performance of the European private equity and venture capital industry has never been better, and 2006 clearly was a record year. For the first time, European private equity fundraising reached €112 billion, an increase by 57% compared to the previous year, with investments in the range of €71 billion – or a 50% rise compared to 2005 – into more than 7500 companies (Marchart, 2007). Figure 1.2 shows the private equity and venture capital investment

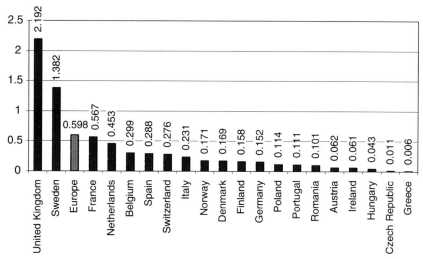

Figure 1.2 Private equity and venture capital investment in Europe, percentage share of the 2006 growth domestic product (European Venture Capital Association, 2007b).

proportion as a percentage of the 2006 growth domestic product (GDP) for the different European countries. Two countries were far ahead of the average, namely UK and Sweden.

1.1.5
The Future of Private Equity and Venture Capital in Europe

Many financial experts believe that a new era has just begun. Not only are today's investors fewer, smarter, more disciplined, and have a more global outlook, they also work with smarter entrepreneurs, many of whom are serial entrepreneurs. According to the German-based Center of Private Equity Research (CEPRES; www.centerofprivateequityresearch.de), better investment selection combined with a strong economic backdrop have led to the highest returns ever.

In addition, according to a recent survey done by the European Venture Capital Association (EVCA; www.evca.com), tax policies and legal codes appear to have become more favorable to private equity and venture capital in most European countries. These changes are reflected impressively in the Private Equity Performance Index (CepreX) provided by CEPRES. CepreX derives from the performance of several hundreds of individual transactions. The partial index for venture capital increased significantly in the past few years and in September 2006 reached a level of more than 26% higher than in the year 2000, the pinnacle of venture before its post-bubble nosedive (Figure 1.3; CEPRES; Romaine, 2007). The future of the European venture capital industry also looks bright, thanks to the mix of low capital volume in the market, a booming economy, and a steep learning curve for venture capital managers, and the European venture market place is expected to continue its rise (Herzog, 2007).

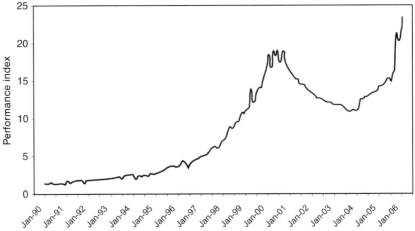

Figure 1.3 European venture performance index CepreX from 1990 to 2006 (CEPRES; Romaine, 2007). The figure shows the gross development of private equity transactions. Fund fees are not included. Therefore, the index development shows the annual return of real deals (company transactions) and not net to the limited partners.

1.2
Development of the Biotech Industry

1.2.1
Definition

By definition, 'biotech' companies are companies whose primary commercial activity relies on the application of biological organisms, systems, or processes or on the provision of specialist services to facilitate the understanding thereof (Hodgson, 2006). The term 'biotechnology' means any technological application that uses biological systems, living organisms, or derivatives thereof to make or modify products or processes for specific use. Biotechnology has applications in four major industrial areas, i.e., health care, crop production and agriculture, nonfood uses of crops and other products (e.g., biodegradable plastics, vegetable oil, biofuels), and environmental uses.

A series of derived terms have been coined to identify several branches of biotechnology, most importantly the white, or industrial biotechnology, focusing on the use of enzymes as industrial catalysts to either produce valuable chemicals or destroy hazardous chemicals, and red biotechnology as applied to medical processes, such as the design of organisms to produce antibiotics or the engineering of genetic cures through genomic manipulation (European Association of Bioindustries; www.europabio.org).

The development of new drug compounds is a complex, long, and costly process. Figure 1.4 describes the product development cycle for a typical new chemical entity

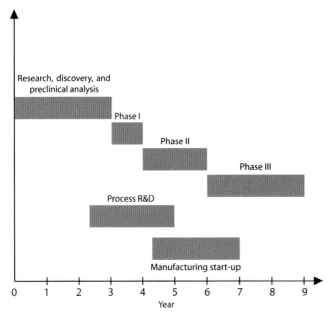

Figure 1.4 The product development cycle of a new chemical entity (Pisano, 1997).

(NCE), starting at the time the entity is discovered. The discovery process leading up to this point can take a number of years. For biotechnology products, or new biological entities, the drug development process is similar to that of NCEs – except that the process R&D and manufacturing start-up begin earlier – with the product development cycle of a new biotechnology-based drug starting when the genes for a specific protein molecule have been cloned (Pisano, 1997).

Once the chemical lead has been discovered or the biological developed, the preclinical research phase begins, with the target to obtain information about the molecule's safety and therapeutic properties. The main goals of preclinical studies are to determine a drug's pharmacodynamics, pharmacokinetics, and toxicity. This stage typically involves subjecting it to a series of screens or tests in both test tubes and laboratory animals. After the company has achieved reasonable confidence in a compound's safety and therapeutic benefits, the next major stage of drug development is human clinical trials. In Phase I clinical trials, the drug is administered to a small sample of healthy volunteers. These trials are designed to determine the drug's safety. Phase I clinical testing also seeks to find out how the pharmaceutical is absorbed and distributed, how long it is active in the body, how it is metabolized by the body, and how it is excreted. The conclusion of Phase I testing leads to Phase II testing, whose primary purpose is to determine whether the drug works, i.e., the drug's pharmaceutical efficacy. Phase II trials also determine the appropriate dosage regime (such as 2.5 mg twice per day, or 5.0 mg once per day) and the form of the drug (such as tablet, capsule, or liquid). In general, the end of Phase II trials marks an important project milestone (Pisano, 1997).

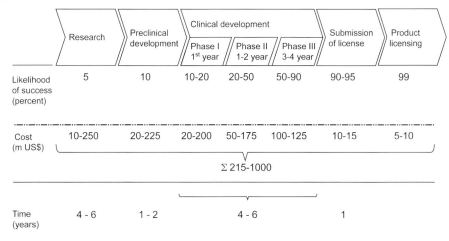

	Research	Preclinical development	Clinical development			Submission of license	Product licensing
			Phase I 1st year	Phase II 1-2 year	Phase III 3-4 year		
Likelihood of success (percent)	5	10	10-20	20-50	50-90	90-95	99
Cost (m US$)	10-250	20-225	20-200	50-175	100-125	10-15	5-10

Σ 215-1000

| Time (years) | 4 - 6 | 1 - 2 | 4 - 6 | | | 1 | |

Figure 1.5 Biotech products take time, require significant investments, and carry considerable risk (US$ million; Dewasthaly, 2007).

Companies review the data obtained in Phase I and Phase II studies very carefully before they make a 'go or no-go' decision on Phase III. Phase III trials, which involve head-to-head comparisons of the new drug against placebo or existing drugs in a large sample of patients, are by far the most costly phase of human clinical trials. It is important to bear in mind that ultimate commercialization is by no means assured simply because a company reaches Phase III clinical trials: approximately 13% of new chemical entities are abandoned after the start of Phase III trials. After successful completion of Phase III trials, the company submits its clinical data for review to authorities like EMEA in Europe and FDA in the US (Pisano, 1997). The entire product development time from research to product licensing ranges over 10–15 years, with associated costs ranging from US$215 million to US$1 billion (Figure 1.5; Dewasthaly, 2007).

1.2.2
Historical Background

In the late 1950s, the biotech industry did not exist, because the key scientific breakthroughs ultimately responsible for its emergence had yet to occur. It still took more than two decades until the first biotech company, Genentech, was founded in the US. The basis for this foundation was the successful invention of a technique for manipulating the genetic structure of cells to synthesize specific proteins, developed in 1973 by Herbert Cohen and Stanley Boyer of the University of California. Genetic engineering first enabled the development of a wide range of proteins into thera-peutic drugs. Within months of Cohen and Boyer's invention, Genentech was formed by Robert Swanson as young venture capitalist and Boyer as co-founder (Pisano, 1997).

In 1978, Genentech closed an agreement with the American pharmaceutical company Eli Lilly, granting Lilly the rights to manufacture and market recombinant insulin for diabetes treatment. In return, Lilly agreed to finance development of the product and to pay Genentech sales-dependent licensing fees. This agreement removed one of the biggest barriers for newly founded companies eager to step into the pharmaceutical business: the large amounts of money needed to finance the expensive R&D development time (Pisano, 2006).

The year 1980 witnessed the public listing of Genentech, and in 1982, the US Food and Drug Administration (FDA) approved insulin for marketing. Many other companies followed the successful example of Genentech, such as Amgen, Biogen Idec, Chiron (now Novartis), and Gilead Sciences (Pisano, 2006). Off to a slower start, the biotech industry in Europe started their attempt to catch up with US biotech developments in the 1990s, with companies such as Serono (now Merck Serono, Switzerland) and Union Chimique Belge (UCB, Belgium) among them. Scientific landmarks that accompanied these developments included the discovery of restriction enzymes, the first transfer of genetic material, and the development of early DNA sequencing methods in the early to mid-1970s, complemented by the invention of polymerase chain reaction (PCR) in 1983 and the completion of the genome sequences for organisms such as drosophila, mouse, and ultimately, humans (Küpper, 2006).

The progress of the industry is reflected by recombinant biological products meanwhile having established markets with multibillion dollar sales, such as erythropoietin, insulin, interferons, monoclonal antibodies, and human growth factors. To date, more than 155 biotech drugs and vaccines have been approved by the FDA and over 370 biotech products and vaccines are currently in clinical testing, targeting more than 200 diseases (Küpper, 2006).

The area of biotechnology has developed into a large, research-intensive industry with more than 4000 companies worldwide and parallel financing and venture capital funding to match. As the largest biopharmaceutical market in the world, the US has benefited greatly from the latest drugs and advances to emerge from biotechnology. In 2006, more than 300 public biotechnology companies in the US were employing over 130 000 people and represented about US$400 billion in market capitalization (Table 1.1; Ernst & Young, 2007).

Table 1.1 Global biotechnology at a glance in 2006 (Ernst & Young, 2007).

	Global	US	Europe	Canada	Asia-Pacific
Public company data					
Revenues (US$ million)	73 478	55 458	11 489	3242	3289
R&D expenses (US$ million)	27 782	22 865	3631	885	401
Net loss (US$ million)	5446	3466	1125	524	331
Number of employees	190 500	130 600	39 740	7190	12 970
Number of companies					
Public companies	710	336	156	82	136
Public and private companies	4275	1452	1621	465	737

Table 1.2 Pipeline of European biotech companies listed on stock exchanges by clinical development phase in 2006 (Ernst & Young, 2007).

Country	Preclinical	Phase I	Phase II	Phase III	Total
UK	99	31	79	37	246
Switzerland	39	14	21	23	97
Germany	37	17	16	7	77
Denmark	28	18	26	5	77
France	17	10	12	3	42
Sweden	14	11	6	5	36
Norway	6	5	5	1	17
Israel	5	3	6	2	16
Belgium	10	1	5	–	16
Ireland	5	3	6	1	15
Italy	6	1	5	2	14
Austria	9	3	1	1	14
Netherlands	4	6	2	1	13
Finland	4	–	1	2	7
Iceland	–	1	4	–	5
Total	283	124	195	90	692

In Europe, the general slowdown in the global economy between 2001 and 2004 also reflected negatively on the European biotech industry. Back on track in 2005, the year 2006 can be qualified as a year where Europe regained its momentum and displayed newly found strengths on every front.

First, the European biotech industry succeeded in growing their pipelines and leading a considerable amount of new chemical and biological entities into the next development phase. Tables 1.2 and 1.3 provide an overview of the pipelines of listed and unlisted European biotech companies by clinical development phase in 2006.

Second, the European biotech industry managed to attract new investors and to show solid growth of key financial metrics, such as a revenue growth of 13% between 2005 and 2006 (Table 1.4). Research and development (R&D) expenses continued to increase and kept pace with rapidly growing revenues as companies realized that R&D investments remain critical to generating value in the long run (Ernst & Young, 2007).

1.2.3
Financing Biotech Start-ups

Each development phase of a start-up company is financed by specific types of investors. In the preseed or seed phases, financing is usually accomplished by sources including the 'three Fs,' i.e., 'family, friends, and fools', as well as government and EU funds. Further on, business angels, and more frequently early-stage venture capitalists continue investing in the start-up. In later development stages, strategic

Table 1.3 Pipeline of European biotech companies not listed on stock exchanges by clinical development phase in 2006 (Ernst & Young, 2007).

Country	Preclinical	Phase I	Phase II	Phase III	Total
Germany	69	21	52	9	151
UK	40	33	30	3	106
France	51	13	19	2	85
Denmark	36	20	16	2	74
Switzerland	30	14	18	7	69
Israel	33	14	18	2	67
Sweden	29	4	14	2	49
Italy	28	10	8	1	47
Austria	30	5	6	4	45
Spain	19	6	7	5	37
Netherlands	15	2	6	1	24
Belgium	13	3	3	5	24
Norway	4	–	9	–	13
Ireland	6	–	5	–	11
Total	403	145	211	43	802

alliances with corporate companies become essential, as are deals with late-stage venture capitalists.

Important liquidity events for already matured companies are initial public offerings (IPOs) and secondary public offerings. An IPO is the first sale of stock by a private company to the public. One important reason for why a private company may wish to go public is to raise capital. A secondary public offering, usually done within a few years following an IPO, is an excellent method for a company to raise additional working capital. There are two main types of secondary public offerings. Thus, an issuer offering involves the issuance of new stock, diluting the ownership position of stockholders who own shares that were issued in the IPO. Alternatively,

Table 1.4 European biotechnology in 2006 and 2005 (Ernst & Young, 2007).

	Public companies			Industry total		
	2006	2005	Change	2006	2005	Change
Financial values (€ million)						
Revenues	9150	7993	14%	13 307	11 765	13%
R&D expenses	2892	2559	13%	5695	5259	8%
Net loss	876	1395	–37%	2541	3280	–23%
Market capitalization	62 165	43 374	43%	–	–	–
Industrial data						
Number of companies	156	122	28%	1621	1613	0%
Employees	39 740	34 250	16%	75 810	68 440	11%

Table 1.5 Financing of biotech companies: US vs Europe in 2005 and 2006 (US$ million; Ernst & Young, 2007).

Type	2006		2005		Change	
	US	Europe	US	Europe	US	Europe
IPO	944	907	626	691	51%	31%
Follow-on and other offerings	16 067	3069	10 740	1577	50%	95%
Venture financing	3302	1907	3328	1738	−1%	10%
Total	20 313	5883	14 694	4006	38%	47%

one or more major stockholders in a company may sell all or a large portion of their holdings. Because no new shares are released, the owners' holdings are not diluted.

In 2006, investors committed a total of more than US$26 billion in the biotech industry, an upswing of 42% compared to previous year (Table 1.5). Venture capital for the first time reached an alltime high of US$5.2 billion. Overall, capital raised increased by 38% in the US and by 47% in Europe (Ernst & Young, 2007).

In Europe, all of these elements, i.e., IPOs, follow-on offerings, and venture financing, contributed to a very strong year 2006, with a grand total of US$5.9 billion, or approximately €4.7 billion, for the entire industry (Table 1.5). By comparison, the increase between 2004 and 2005 was a mere 18%. A total of 32 companies went public in 2006. While the importance of IPOs and follow-on financing vehicles increased, venture capital investments did not hold path to the same degree. Even so, the overall amount of venture capital raised by European biotech companies in 2006 set an alltime record, passing US$1.9 billion, or approximately €1.5 billion, for the first time ever (Table 1.5, Figure 1.6; Ernst & Young, 2007).

Overall, venture capital expenditures differed greatly between countries. While France and some smaller countries, such as Belgium, Spain, and Austria,

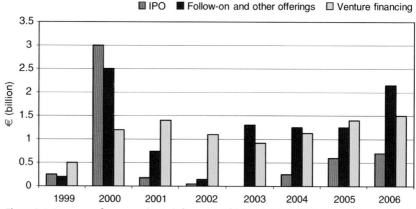

Figure 1.6 Summary of European biotech financing (€ billion), 1999–2006 (Ernst & Young, 2007).

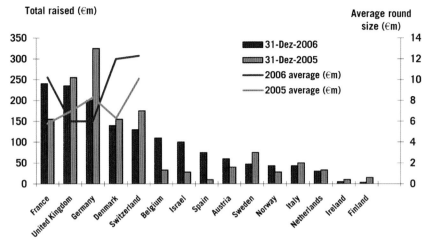

Figure 1.7 European venture capital by country (€ million), 2006 and 2005 (Ernst & Young, 2007).

significantly increased their venture capital volume, others, including Germany and Switzerland, experienced declines relative to 2005. In Germany, the reason was the preponderance of early-stage deals with lower average round sizes, while in Switzerland, the decline resulted from 2005 having been an exceptionally strong year, in part because of two pre-IPO rounds raised by Speedel, totaling €76 million (Figure 1.7; Ernst & Young, 2007).

With regard to IPOs, the industry's market capitalization took a big step forward in 2006 (Figure 1.8), with a 43% upswing compared to the previous year (Table 1.4; Ernst & Young, 2007). Public investors who in previous years had invested elsewhere, began to regain trust in biotech and even supported public offerings of relatively young companies.

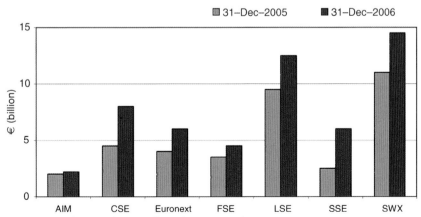

Figure 1.8 Market capitalization of biotech companies by main European stock markets (€ billion; Ernst & Young, 2007).

Total market capitalization of the IPO class was €3.5 billion in 2006, almost identical to 2005. Fundraising success varied from exchange to exchange. The Swiss exchange (SWX) in Zurich proved outstanding both in terms of capital raised and average deal size. Surprisingly, the London Stock Exchange (LSE) and the Frankfurt Stock Exchange (FSE) were less successful. With a total of eight IPOs and €107 million raised, the Alternative Investment Market (AIM) in London attracted the highest number of deals and achieved a similar ranking as the stock market in Copenhagen (CSE) and Euronext. AIM was generally used by smaller enterprises with lower capital needs (Figure 1.8; Ernst & Young, 2007).

1.2.4
Exit Routes in the Biotech Industry

From the beginning, investors are generally motivated by a clear idea of the exit strategy of a young company and the method through which they can cash in on their investment. During their long development times, most biotech start-ups try to set up parallel exit options, ultimately enabling them, together with their investors, to pick the best strategy to realize the inherent value of the company. In the biotech industry, the most frequent exits are trade sales and IPOs.

For an IPO to be successful, the company first has to invest enormous amounts of time and money into building a pipeline that is likely to attract investors. For many European biotech companies, the price of establishing such a pipeline is too high relative to what can be raised on the public markets. Moreover, having gone public, young biotech start-ups will be challenged by a much higher visibility and receive increased scrutiny from both the media and financial analysts, particularly if the first products have not yet reached the market.

Therefore, investors and companies are looking at trade sales as an alternative route (Ernst & Young, 2007). Acquisition activity has recently heated up, and the reasons for this are manifold. The lackluster IPO market, the high price of licensing deals that have made outright acquisitions more attractive, an opportunity for pharmaceutical companies to repatriate foreign earnings, new requirements on public companies in the form of Sarbanes–Oxley, and new restrictions on analysts all work together to fuel this trend (Levine, 2005).

Finally, another exit strategy is the MBO, i.e., managers or executives of a company purchase the controlling interest in a company from existing shareholders. This is usually not a viable option for biotech companies, because their entrepreneurs do not dispose over such high capital needs.

In the first six months of 2007, an amazing 37 biotech firms went public worldwide, compared to 50 IPOs in the whole of 2006. Most of the newly listed companies were based in North America and Europe. North America had 15 companies involved in IPOs and Europe had 14, overshadowing China with four, Israel with three, and Australia with one (Scrip, 2007b).

Moreover, in 2006, the European biotech industry continued to engage in mergers, acquisitions, and alliances. While the number of M&A transactions was flat, many individual transactions were interesting with regard to valuations and targets. In a global trend, pharmaceutical companies around the world invested large amounts of money in biotech companies with promising platforms in areas such as antibodies or vaccines, as they sought to buy what they hoped could become the engines for new generations of cutting-edge products. On the alliance side, there has been a recent shift to early-stage deals with companies covering broad applications through innovative platforms or individual products with treatment options in several indications (Ernst & Young, 2007).

1.2.5
The Challenges to European Biotech

Today, the European biotech industry accounts for close to 75 000 jobs and over 1600 companies. While the expansion of the EU has brought significant new opportunities to the European biotech industry, including cost advantages, it has also increased the legal and regulatory complexity of developing a drug candidate through to approval by the European Medicines Agency (EMEA; Ernst & Young, 2007).

Moreover, as indicated earlier, it has become tougher for companies to raise venture capital. The reasons for this are manifold. For one thing, consolidation has thinned the ranks of venture capital firms. Also, the vibrant European IPO market, swallowing large amounts of capital, may have reflected negatively on the venture capital environment. The increasing complexity of venture capital consortia, dilution and liquidation preference issues, and valuation discrepancies have further aggravated the situation. Another challenge has to do with the increasing unwillingness of venture capitalists to fund discovery or early stages of clinical development. The companies suffering most from this funding gap are in critical development phases, where product candidates have emerged but financial support for proof-of-concept clinical studies is lacking. At the same time, proof-of-concept is the most important decision factor for successful investment decisions.

While aggregate fundraising was high, the number of rounds fell and average capital raised per round increased. More money seems to have been made available, but it went to fewer companies. With the venture capital focus on late-stage companies (Figure 1.9), the question is how young biotech start-ups will attract the capital needed, and this will increase the overall challenge on the industry's sustainability (Ernst & Young, 2007).

1.2.6
The Future of the European Biotech Industry

Next to private equity and venture capital fund raising, additional opportunities for entrepreneurs seem to be developing, especially through newly installed EU public policy pathways. In March 2000, the Lisbon Agenda was adopted, with the goal to make Europe the "most competitive and the most dynamic knowledge-based economy in the world by 2010" (European Commission; http://ec.europa.eu/growthandjobs/index_en.htm). To achieve the Lisbon objectives and to encourage R&D, the EU initiated the Seventh Framework Program for Research and Technical

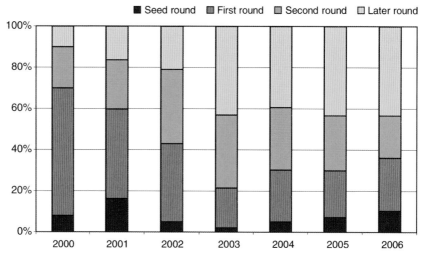

■ Seed round ▒ First round ▨ Second round ☐ Later round

Figure 1.9 European venture funding by round class, 2000–2006 (Ernst & Young, 2007).

Development (FP7) in funding food, agriculture, and biotechnology research, with an overall budget of €53.2 billion from 2007 through 2013 (Figure 1.10; Community Research & Development Information Service; http://cordis.europa.eu). This budget was divided up among the following areas by an indicative breakdown:

• Collaborative research ('cooperation'), supporting all types of research activities carried out by research bodies in transnational cooperation targeted to gain or consolidate leadership in key scientific and technology areas

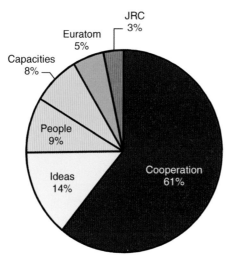

Figure 1.10 Indicative breakdown of the FP7 (http://cordis.europa.eu).

- Frontier research ('ideas') within the framework of activities commonly understood as 'basic research,' targeted to produce new knowledge leading to future applications and markets

- Marie Curie Actions ('people'), a program dedicated to stimulating researchers' career development, mobility, and training

- Research capacity ('capacities'), a program complementary to the cooperation program aiming to enhance research and innovation capacities throughout Europe as well as their optimal use (e.g., support for the coherent development of policies)

- EURATOM, carrying out energy research activities for nuclear research as well as training activities

- Joint Research Centre (JRC), providing customer-driven scientific and technical support to the conception, development, implementation, and monitoring of EU policies. Priorities include competitiveness and innovation, supporting the European Research Area (ERA), research in the areas of renewable and cleaner energies and transport, life sciences, and biotechnology.

The collaborative research program, called 'cooperation,' with an overall budget of €32.4 billion, is supposed to bring together the best talents from across Europe to tackle areas such as health, food, agriculture, biotechnology, energy, and the environment. Figure 1.11 provides a detailed overview of spending per area.

Since their launch in 1984, the Framework Programs have played a lead role in multi-disciplinary research and cooperative activities in Europe and beyond. FP7

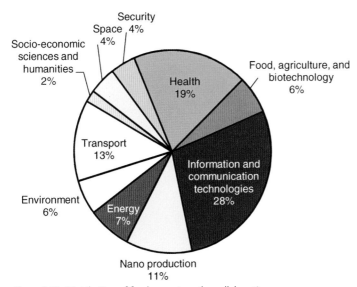

Figure 1.11 Distribution of funds spent on the collaborative research program (http://cordis.europa.eu).

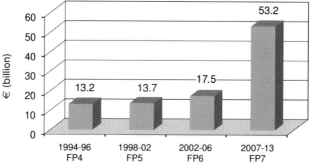

Figure 1.12 Evolution of EU research framework program budgets (€ billion; http://cordis.europa.eu).

continues this task and is also both larger and more comprehensive than earlier Framework Programs. FP7 reflects the largest funding allocation yet, representing an enormous increase in budget compared to FP6 (Figure 1.12).

Participation in FP7 is open to a wide range of organizations and individuals, including universities, research centers, multinational corporations, small to medium-sized enterprises, public administrations, and even individuals.

Beyond FP7, there seem to be continuous and increasing efforts within the EU to set up better frameworks and incentives that use state aid as an instrument to boost research, development, and innovation. One case in point is the new Research, Development and Innovation (R&D&I) Framework, which took effect on 1 January 2007 and set out a series of guidelines for specific types of state aid measure – such as aid for R&D projects, aid to young innovative enterprises, and aid to innovation clusters – that could encourage additional R&D&I investments by private firms, thus stimulating growth and employment and improving Europe's competitiveness (Ernst & Young, 2007).

Overall, the most encouraging news in Europe in 2006 was related to the field of product development, and this is expected to continue in 2007 and beyond. With wider and more mature pipelines, European companies should be well poised to bring innovative new products to the market. Ultimately, the ability to deliver on that promise will become the true determinant of sustainability (Ernst & Young, 2007).

2
Methodology

2.1
Formulating the Research Goal

2.1.1
The Entrepreneur's Dilemma

While there are many definitions of the term 'entrepreneurship,' entrepreneur Mark J. Dollinger describes entrepreneurship as "the creation of an innovative economic organization for gain or growth under conditions of risk and uncertainty" (Dollinger, 2003). Some 30 years earlier, both Frank H. Knight and Peter Drucker had also clearly stated that entrepreneurship was about taking risk (Drucker, 1970; Knight, 1967). The behavior of the entrepreneur reflects his willingness to put his career and financial security on the line and take risks in the name of an idea, spending much time as well as capital on an uncertain venture.

The biotech entrepreneur often has a high-level technical and scientific background. It is quite common for scientists who have spent a number of years in an academic environment to start a new company based on a technology platform or a discovery they made during their tenure at a university or research institute. Alternatively, a scientist or a group of scientists from a rather large company may decide to leave and start their own venture based on some of the R&D work they had either performed or thought about doing for their previous employer. This type of business may or may not be a spin-off from the parent company. In some cases, business persons may join the scientists in the start-up of the new company. A less frequent but still relatively common phenomenon is the start-up of a new venture by someone with a stronger background in business than in science.

What all these different types of start-ups have in common is that, at the beginning there is a visionary entrepreneur who is convinced of his discovery and prepared to step out of a 'protected' environment to take a significant amount of risk in creating his own business. After founding the start-up and continuing through the entire development process, the most critical element for the entrepreneur will be to identify adequate sources of funding to support his expensive R&D research. In this search for funding, the entrepreneur will, sooner or later, come in contact with

Biotech Funding Trends: Insights from Entrepreneurs and Investors.
Alexandra Carina Gruber
Copyright © 2009 WILEY-VCH Verlag GmbH & Co. KGaA, Weinheim
ISBN: 978-3-527-32435-4

venture capitalists providing funding for new businesses – but at the same time taking away some share in future profits and 'diluting' the entrepreneur. In many cases, there will not be too many alternative options of funding available to the entrepreneur. He may not be able to choose between the individual financing models and their advantages and drawbacks but, rather, may have to agree to what is offered to him. Down the road, the entrepreneur may be rewarded with a strategic alliance with a pharmaceutical company and still later with the trade sale of his company. But until this time has come, he has to pass through many years of endurance, persistence, and hard work and has to successfully mitigate the risk in all the different development stages (Pappas, 2002; Robbins-Roth, 2001).

Business opportunities for such high-growth biotech start-up companies are truly entrepreneurial and over decades have contributed a lot to the modernization of our economy. Nevertheless, entrepreneurship has remained a difficult undertaking, and a vast majority of new businesses fail. This sets entrepreneurship apart from strategic management. Importantly, entrepreneurship is a critical subject to be integrated in university curricula. The goal is to make students from diverse disciplines familiar with the principles of starting their own business in a high-growth start-up venture, with teaching modules ranging from available financing options and networking trends to the opportunities and threats involved in starting a business.

2.1.2
The Investor's Dilemma

In general, the investor will be concerned about the dual relationship between risk and return. In the early phases of a start-up, the risk for investors is extremely high. The picture is particularly complex for biotechnology companies, considering their long development times, the high capital requirements, and the high risk involved in each individual development step from preclinical to clinical Phase III studies. In any of these development phases, elaborate and expensive studies may fail to achieve their endpoints, potentially leading to premature closure of the entire development program (Figure 2.1).

In the early days of a biotech start-up involved in preclinical research, the risk for investors is definitely at its peak, but so is the expected return if the company succeeds. The more advanced the development process, the lower is the risk associated with the investment in the start-up, but the lower will also be the expected return on the investment. It is therefore up to the individual investor to decide whether to step in either early or later in the development of a biotech company and to balance risk and return on behalf of his customers, the limited partners (LP) who provide the capital (Frei, 2007).

2.1.3
Financing Models

Financing biotech start-ups is definitely a challenge for any investor who decides to become part of the financing structure of a young company, regardless of his

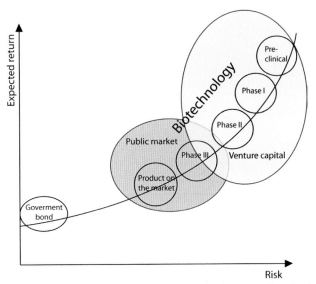

Figure 2.1 Relationship between risk and expected return in drug development (Frei, 2007).

risk-taking capability. The central research question of this study was based on the general financing model matrix for innovative start-ups. Thus, each development phase of a start-up company is usually financed by specific types of investors. Figure 2.2 gives a schematic presentation of the degree of involvement of the

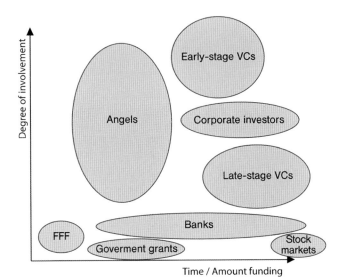

Figure 2.2 Degree of involvement and funding provided by individual types of investors (Hellmann, 2006).

individual investors in the development of a company and the amounts of funding they provide (Hellmann, 2006).

In the preseed and seed phases, financing in many cases starts with the 'three Fs' – family, friends, and fools. After that, it is usually the government that steps in with grants and loans, followed by business angels and early-stage venture capitalists. In the later stages of development, strategic alliances with corporate companies are key to success, as are deals with late-stage venture capitalists or banks (Hellmann, 2006). In general, much higher amounts of funding will be needed in later development stages. The last link in the financing chain of a biotech start-up is the exit, usually in form of an IPO, trade sale, or MBO (Metrick, 2007).

Based on the financing matrix given in Figure 2.2 and against the backdrop of the complexity of biotech development, the primary question was whether a biotech start-up went through the same development phases and was financed by the same types of investors as other innovative start-up companies, such as IT, industrial products, or services companies. Do all biotech companies share the same financing model? What do investors think about the relative importance and interaction of available financing options. Which financing options are used in early-stage start-ups, and which are used in later stages? What do biotech entrepreneurs and various types of investors consider to be the major advantages and drawbacks of these different financing sources? How can the gap between early and follow-on financing be closed? Are there any new and creative financing approaches available to young biotech companies? What do entrepreneurs and investors think about the different exit scenarios available?

2.1.4
Interpersonal Relationships and Cooperations

Another focus of this project was to determine what the interpersonal relationships and alliances in the field of biotechnology looked like, what the characteristics of the individual key players were, and how they interacted with one another. How can early- and late-stage alliances be set up most successfully and what is expected and required of the individual types of investors in each development phase to best support a young biotech company? How can networking between the individual players be intensified, and how significant are such networks for a biotech start-up? Why have many biotechnology start-ups failed?

2.1.5
Practical Guidelines for Entrepreneurs and Venture Capitalists

Analysis of the available funding and networking options and their relative importance from the perspectives of the major players in this industry finally enabled practical checklists for new entrepreneurs and investors to be developed, giving insight into the lessons learned by their peers who had gone before them and helping improve mutual understanding between the parties involved in the successful launch of a biotechnology company.

2.2
Conducting the Study

To answer the questions outlined above, qualitative and standardized interviews were carried out with three groups of interview partners, i.e., entrepreneurs, venture capitalists, and investors other than venture capitalists. The third group of other investors was considered a control group of sorts, because they are usually good observers of the relationship between the entrepreneur and the venture capitalist.

2.2.1
Qualitative Interviews

Qualitative interviews were used because they are most useful for the evaluation of programs aimed at individualized outcomes and for exploring individual differences between participants' experiences and beliefs (Sewell, 1998). They usually provide the following advantages compared to quantitative surveys. By using open-ended questions, qualitative interviews provide the opportunity for program participants to describe their experiences in their own words (Froschauer and Lueger, 2003; Patton, 1987). Also, by listening to the interviewees' perspective, it is possible to find out about important aspects that might not be commonly known or might not have been considered in a quantitative survey. Important aspects may include the terminology that participants use to describe their experience, attitudes, or behaviors. In this respect, it is essential for the interviewer to be consistent, i.e., to stick to asking the same questions in the same order. Also, the interviewer must take a neutral stance, e.g., by avoiding agreeing or disagreeing with a participant, avoiding suggesting an answer, or interpreting a question for a participant (Kiernan et al., 2003).

Qualitative interviewers aim to go below the surface of the topic being discussed, explore what people say in as much detail as possible, and uncover new ideas that were not anticipated at the outset of the research (Britten, 1995). Questions based on behavior or experience, opinion or value, feeling, knowledge, or sensory experience and those asking about demographic or background details are usually most suitable for qualitative research.

The three customized questionnaires used in the study, one for each of the three groups of interview partners, were divided into subsections dealing with the most important attributes of entrepreneurs and investors, the financing models applied, and the experiences with alliances and networks in the early or late stages of a start-up company. In each section of the questionnaire, special emphasis was placed on the specific characteristics, challenges, and opportunities that set the biotechnology industry apart from other areas, such as IT or telecommunications.

2.2.2
Selection of Interview Partners

Sampling strategies are generally determined by the purpose of the research project. Statistical representativeness is not normally sought in qualitative research. Similarly, sample sizes are not determined by hard-and-fast rules, but by other factors, such as the depth and duration of the interview and what is feasible for a single interviewer. The idea of this type of sampling is not to generalize to the whole population, but to indicate common links or categories shared between the groups. Informants are identified because they will enable exploration of a particular aspect of behavior relevant to the research. This approach to sampling allows the researcher to include a wide range of types of informants and to select key informants with access to important sources of knowledge. Large qualitative studies do not often interview more than 50 people (Mays and Pope, 1995a, b).

In this study, the selection of interviewees was based primarily on internet research, recommendations from university, and referrals from the interview partners themselves. The intention was to obtain an approximately equal number of interviews with biotech entrepreneurs, venture capitalists, and other investors. The sample size of about 15 per group was largely determined by the number of European venture capitalists being rather limited, particularly when counting only those with access to 'fresh capital' or managing funds with considerable capital reserves. The estimate goes that there are no more than approximately 50 of such funds in Europe.

For the group of entrepreneurs, the following selection criteria were predefined:

- Long-term experience in the field of biotech
- International experience and mindset, i.e., having pursued at least a part of their educational or professional careers in countries abroad
- A balance of entrepreneurs working in different research areas, including not only those working on new pharmaceutical substances, but also those involved in contract research, platform technologies, medical technology, or diagnostics
- A mix of entrepreneurs who developed their company as a spin-off from either an academic institution or another pharmaceutical company
- A mix of entrepreneurs in different development stages, i.e., seed, early, and late stages
- A mix of entrepreneurs having received venture capital and those who had not
- A good mix of first-time and serial entrepreneurs
- A balance between founders who still fulfilled the role of CEO and those who had already taken an external CEO on board.

For the group of venture capitalists, the selection was based on the following criteria:

- A good balance of experts from well known European venture capital companies having access to 'fresh capital' or managing funds of a considerable size
- A mix of venture capitalists from different European countries
- A mix of venture capitalists working in seed, early, or late-stage financing
- Specialization in biotech funding
- Long-term experience in venture capital funding.

For the group of investors other than venture capitalists, the selection criteria were:

- A mix of investors, i.e., those who either worked in seed financing, such as government institutions, were involved in late-stage financing, or focused on IPOs or trade sales, such as banks or financial advisors
- A mix of investors from different European countries
- Special knowledge in the biotech field
- Long-term experience in private equity, venture capital, or government funding.

All interviews were initiated by a formal invitation per email. A total of 50 invitations were sent out. The interest in participating in the interviews was overwhelming, with 88% of all those invited spontaneously agreeing to participate. Interviews were conducted between April and July 2007. Most of the interviews took place in personal face-to-face meetings, guiding the experts through a set of between 23 and 25 questions depending on the group of interview partners. Approximately 30% of interviews were conducted by phone due to the distant location of the interview partner. During the interviews, each lasting 60–120 minutes, notes on the answers to each question were taken. Most of the phone interviews with interview partners from abroad were conducted in English, whereas most of the personal interviews were conducted in German.

2.3
Evaluating the Results

Qualitative data and their transcripts produce a large volume of material that must be condensed, categorized, or otherwise interpreted and made meaningful (Sewell, 1998). Qualitative research results can be evaluated in several ways. One available method is what is referred to as the 'socio-scientific paraphrase,' a technique attempting to use hermeneutics to understand and interpret the subjective perspective of its subjects (Department of Informatics, Duisburg University Essen; www.is.informatik.uni-duisburg.de).

According to existential hermeneutics, a human being is not an isolated inquirer attempting to reach others or the outside world from his encapsulated mind, but is already sharing the world with others. Within this world of open possibilities, a person meets others and shares things and beliefs, developing in its variety and complexity. Hermeneutically, the process of storage and retrieval of information has to do with the relationship between the existential world-openness of the inquirer, his different open and socially shared horizons of pre-understanding, and the established horizon of the system. The information-seeking process is basically an interpretation process that is closely related to the life context and the background of the inquirer (Capurro, 2000).

The hermeneutic circle describes the process of understanding a text hermeneutically. It deals with the central ideas that one's understanding of the text as a whole is established by reference to the individual parts and one's understanding of each individual part by reference to the whole. Neither the whole text nor any individual part can be understood without reference to one another, and hence, a circle develops. With the statedness of a part of a community background in a system, the inquirer tries to match his questions and backgrounds of pre-understanding against it. The dialogue is a specific form of the 'hermeneutic circle.' Questions arise within the dialogue partners' pre-understanding, which is itself the output of having asked questions, and so on. This attitude is existentially grounded in the fact that we are embedded in an already structured world, which depends on factors such as historical situation, culture, or language, while at the same time being confronted with a wide range of possibilities (Capurro, 2000).

In qualitative research, descriptions alone cannot provide explanations. Therefore, data need to be sifted and decoded to make sense of the situation, events, and interactions observed, an analytical process usually starting during the data collection phase (Mays and Pope, 1995b). On the day of the interview, the notes taken were reviewed and transferred into a Microsoft Excel assessment matrix. The interview results from each of the three groups were first evaluated separately and then compared between groups to identify similarities and differences.

The interpretative procedures were defined before the analysis. Thus, a coding frame was developed for each set of questions to characterize each utterance. Based on this coding scheme, the data obtained were classified and categorized. For some questions, a combination of qualitative and quantitative analysis was applied. However, the overall approach to the analysis remained qualitative. This systematic approach of research design, data collection, hermeneutic interpretation, and presentation of results was chosen to guarantee the validity of the study.

3
Data Analysis and Interpretation of Results

3.1
Description of Interview Partners

3.1.1
Entrepreneurs

Fourteen entrepreneurs, primarily from German-speaking countries ($n = 13$), were interviewed. Seven of the interview partners came from or were working in Austria, five in Germany, one in Switzerland, and one in the United Kingdom. Eight entrepreneurs were the original founders of their respective companies and held the position of CSO or CEO at the time of the interview. Two of these were serial entrepreneurs. Six entrepreneurs had joined already existing companies in their current roles as CEO or CFO. The majority of companies were university spin-offs, followed by company spin-offs.

At the time of the interviews, all entrepreneurs had 10–20 years of experience either from the pharmaceutical or biotech industry or from the venture capital field. All entrepreneurs had been living in various parts of the world primarily in Europe or the US, pursuing parts of either their educational or professional careers abroad – a factor contributing greatly to the international mindset the interviewees had developed over time. Nine had an educational background in science and five had a business degree. Three entrepreneurs were involved in preseed, seed, or early research, eight in early and three in late-stage development. Half of the entrepreneurs were focusing on research on new biotechnology-based products (e.g., vaccines) or new chemical entities, the other half was primarily working on the development of new platform technologies or diagnostics. Two entrepreneurs had closed contractual research collaborations in the field of white biotechnology. Ten of the 14 companies had already received venture capital in either a first or second financing found, whereas the remaining four companies were close to their first venture capital financing round.

Biotech Funding Trends: Insights from Entrepreneurs and Investors.
Alexandra Carina Gruber
Copyright © 2009 WILEY-VCH Verlag GmbH & Co. KGaA, Weinheim
ISBN: 978-3-527-32435-4

3.1.2
Venture Capitalists

Fifteen venture capitalists were interviewed. Five of the interviewees came from Germany, three from Austria, three from Switzerland, two from France, one from the United Kingdom, and one from Italy. The interview partners had between 10 years and more than 30 years of experience in the biotech or venture capital industry, and most had a degree in science as well as a high level of operational knowledge. Almost all venture capitalists ($n = 13$) had studied abroad in either Europe or the US before returning to their home country for their current positions. The majority of venture capitalists ($n = 11$) were managing funds focusing on the biotech industry, but also covering IT and other high-growth business areas. At the time of the interview, the interview partners were working for companies advising and managing private equity and venture capital funds with volumes ranging from approximately €40 million to €400 million. Five of the venture capitalists were investing primarily in seed and early-stage projects, whereas ten were investing in early- and late-stage companies.

3.1.3
Other Investors

The third group was the most heterogeneous group of 15 interview partners from Austria ($n = 6$), Germany ($n = 4$), Switzerland ($n = 2$), the UK ($n = 1$), Italy ($n = 1$), and the US ($n = 1$), with jobs ranging from government funding organizations ($n = 5$) and consulting agencies ($n = 6$) to individuals acting on behalf of other investors ($n = 4$), such as pharmaceutical corporations or banks. Most members of this group ($n = 12$) had been pursuing part of their educational or professional career abroad. They all had between 10 and 30 years of experience, with a strong background in either finance or biotech. Three investors worked in preseed, four in seed, four mainly in early-, and four in late-stage financing.

3.2
Major Characteristics of an Entrepreneur

One of the first questions in each of the three questionnaires was related to the strengths and capabilities that entrepreneurs, venture capitalists, and other investors thought were required of an entrepreneur. Multiple answers were allowed. For evaluation purposes, the answers to the open questions were grouped into topical categories. The results are given in Table 3.1.

The major strengths of an entrepreneur were seen similarly across all three groups. Thus, in the eyes of both the entrepreneur and the venture capitalist, the most important skills of the entrepreneur clearly are of a leadership and interpersonal nature. Based on the individual answers given, these skills involve sharing responsibilities, being able to maintain professional partnerships, getting people with complementary skills on board, and working in teams.

Table 3.1 Strengths of an entrepreneur from the point of view of . . .

. . . the entrepreneur (*n*)	. . . the venture capitalist (*n*)	. . . other investors (*n*)
Leadership and interpersonal skills (6)	Leadership and interpersonal skills (9)	Experience (7)
Experience (5)	Experience (9)	Conviction (7)
Scientific and operational knowledge (5)	Scientific and operational knowledge (8)	Leadership and interpersonal skills (6)
Conviction (5)	Conviction (6)	Scientific and operational knowledge (3)
Strong nerves (4)	Strong nerves (5)	Understanding the interests of the venture capitalist (3)
Networking (3) Understanding the interests of the venture capitalist (3)	Networking (4)	Strong nerves (2) Networking (1)

The second most important characteristic of an entrepreneur is personal and professional experience. The entrepreneur is expected to have a proven track record in the industry and to be able to demonstrate leadership both at a technological and a strategic level. Bringing an entrepreneurial spirit to the job is obviously regarded as beneficial, considering the multitude of hurdles that need to be overcome in the first years of a start-up company.

The third major strength required of an entrepreneur is a solid scientific background and operational knowledge of how to successfully run a business. Entrepreneurs must be able to master the balancing act between research-related and business-related activities. The ideal entrepreneur will be an all-in-one genius, combining market- and process-oriented know-how and selling and business development skills with an ability to think in terms of basic research.

The group of other investors considered experience and conviction – a clear vision and strong belief in the company and its products – the most important traits of an entrepreneur. Strong nerves, i.e., showing a high frustration threshold and persistence, as well as networking also played a role.

3.2.1
The Dual Role of the Entrepreneur

One questionnaire item dealt with the generally recognized entrepreneurial challenge to combine two essentially different roles, i.e., that of being a scientist and a general manager at the same time. Again, the results were very similar across all the three groups. It was generally acknowledged that the personality traits of a scientist and those of an entrepreneur are mutually exclusive. However, there have always been exceptions to the rule, and there are some company founders who have successfully grown into the role of a CEO.

For many company founders, however, it will be essential to be able to let go and to dissolve any personal union as early on as possible by clearly separating the functions of CEO, CFO, and CSO. With most founders regarding their company as their lifetime achievement, one challenge investors may have to grapple with is how to give the entrepreneur full support in separating corporate functions. In this context, the interview partners emphasized that it was important to find partners who complemented the founder in terms of skills and experience and to build them into a professional management team. Members of the team have to develop a high level of self-awareness and should be coached about their personal strengths and weaknesses. All functions should work together in defining the company milestones and in achieving them, while at the same time seeing to it that adequate financial support is guaranteed.

Obviously, there are no hard-and-fast rules of how to tackle this challenge. When starting a company, it will often be beneficial for the entrepreneur to acquire management skills to be better able to understand the mechanisms behind a business plan and the interplay of the various forces driving a business, such as market, competition, and pricing. From the very outset, the vision of the entrepreneur should be to establish a business which represents an under-served market need and which is global and sustainable (Lang J and the Cambridge Entrepreneurship Centre, 2001). Later, as the company structure becomes more diversified, complex, and demanding, additional staff with complementary skills will have to come on board to lead the company into the next development phase. Different development stages call for different skill sets to have the company value increase in the most effective and sustainable manner. Failure of any member of the management team to understand this may sometimes require tough personnel decisions to be taken and the management to be reshuffled. Delaying such decisions may put the success of the entire company on the line.

3.3
Major Characteristics of a Venture Capitalist

One questionnaire item was related to the strengths and capabilities that entrepreneurs, venture capitalists, and other investors thought were required of a venture capitalist. The results are given in Table 3.2.

Table 3.2 Strengths of a venture capitalist from the point of view of . . .

. . . the entrepreneur (*n*)	. . . the venture capitalist (*n*)	. . . other investors (*n*)
Networking (6)	Experience (8)	Experience (9)
Capital (5)	Networking (6)	Networking (7)
Knowledge (5)	Capital (5)	Knowledge (7)
Experience (5)	Knowledge (5)	Long-term perspective (4)
Long-term perspective (5)	Long-term perspective (4)	Interpersonal skills (3)
Persistence (1)	Negotiating skills (3)	Capital (2)
	Interpersonal skills (2)	Negotiating skills (1)
	Persistence (1)	

The categories identified by all three groups as the most important strengths required of venture capitalists differed from those identified for entrepreneurs. However, the results regarding the major strengths required of a venture capitalist were again fairly homogenous across groups, with network, experience, and knowledge ranking first. For the entrepreneur, the most important factor mentioned was being able to make use of and profit from the regional or global network of the venture capitalist. Other equally important factors were capital and the resulting access to financial markets, know-how, experience both in the pharmaceutical industry and in deal making, and a long-term perspective. Several entrepreneurs pointed out that one of the major difficulties was that venture capitalists very often had too short-term oriented an outlook. Therefore, many entrepreneurs said they considered it essential to partner with a venture capitalist who knows the biotech/pharmaceutical business and understands that research and development activities take time.

At the same time, entrepreneurs seem to be well aware that it can be difficult to maintain the professional partnership with an investor in case of missed milestones or development failures and that their relationship is limited to a certain period. Some venture capitalists and other investors also mentioned negotiating and interpersonal skills as being important characteristics of a venture capitalist – requirements which appear to be less evident to the entrepreneur.

3.3.1
Global Reach of the Fund

One questionnaire item in this section asked whether the interview partners believed in funds with a purely national, regional, or global appearance. What became clear from the answers given was that in early development phases, at a time when start-ups need a great deal of guidance and support, a local or regional office is preferred. The later the development stage, the more capital will be required and the more global the fund is likely to be. Both the life science industry, an extremely capital-intensive area, and the investment markets are global businesses.

Among entrepreneurs, there was a clear tendency in favor of regional and global funds, although some entrepreneurs, particularly those in early stages, favored national funds or mentioned no preference. Some interviewees acknowledged that one of the advantages of regional funds was that they could provide local public relations and lobbying to attract global funds for the next financing round and that regional funds usually had a much broader experience than their purely national counterparts.

The venture capitalists also favored regional and global funds, and they spelled out their reasons in greater detail. With local funds not considered a viable alternative, the fund had to have at least a regional reach for the following reasons:

- Venture capitalists need to be close to the company and its research for counseling and consulting

- The network between the venture capitalist and the entrepreneur needs to be intensified
- Larger funds stand a greater chance of being able to take the start-up to the next financing round
- The fund must be able to invest in companies outside its home country
- The biotech industry is a global industry
- Capital markets, too, are global, primarily to spread risk.

Some of the venture capitalists opting for funds with a global reach also indicated that global funds needed to have regional offices located close to where the biotech start-up is located to be able to provide practical support and added value for the newly founded company.

The group of other investors was also clearly in favor of funds with a global appearance, with over two-thirds of interviewees favoring global funds. Some investors pointed out that there were two factors making investments in Europe more difficult than they are in the US. The first is the wide variety in cultural idiosyncrasies and languages. Second, European venture capital funds often find themselves in a maze of national administrative, regulatory, and tax regulations still differing considerably between countries and rendering cross-border investment difficult, time-consuming, and costly. As a result, the European venture capital market is highly fragmented along national lines. This issue has been addressed by the EVCA for many years and has now been taken up by the European Commission with a view to improving the regulatory framework to lower operational costs and risks, raise returns, increase the flow of venture capital, and improve the functioning of venture capital markets. Whereas a first step would be to eliminate double taxation within the EU, the final goal is the free movement of capital between member states (Scrip, 2008).

Some interview partners added a critical remark about the prime focus of many funds, stating that the biotech industry was often underrepresented because of the longer development horizons compared to other high-growth start-ups, such as IT or telecommunications. This is why experts in many European countries have been trying hard to introduce dedicated biotech funds. For example, in Austria efforts are currently under way to set up a fund of approximately €1–2 billion to make earmarked money and specific expertise available to biotech start-ups. The capital should be made available through banks, insurance companies, and private equity firms (Zwettler, 2007).

3.4
The Most Important Partnering Criteria

The group of entrepreneurs was asked what they thought the single most important partnering criterion was when looking for a partner investor, and the groups of investors were asked what they looked for in entrepreneurs they would select as partners. The results are given in Table 3.3.

Table 3.3 The most important partnering criterion from the point of view of . . .

. . . the entrepreneur (*n*)	. . . the venture capitalist (*n*)	. . . other investors (*n*)
Experience (5)	Sound science (5)	Good management team (5)
Scientific know-how (3)	Good management team (5)	Uniqueness (4)
Trust (2)	Uniqueness (2)	Sound science (3)
Networking (2)	Credibility (2)	Credibility (1)
Financial resources (2)	Intellectual property situation (1)	Intellectual property situation (1)

3.4.1
The Entrepreneur's Perspective

For most entrepreneurs, the major partnering criterion was the experience of the investor, followed by their scientific knowledge, reliability and trust, network, and financial resources. Interestingly, when entrepreneurs were asked what in their opinion the most important criterion for choosing an investor was, it was not money or the investor's network that ranked first, but experience and knowledge. Thus, the entrepreneur will look for an investor with a proven track record in the industry who has successfully completed similar projects in the past, someone who has been through the entire process, and who has learned the tools of his trade from scratch. The investor must be competent in his particular investment area, have an experienced management team that is capable of anticipating the challenges ahead, and have a clear vision regarding the company's exit strategy. He is expected to have adequate levels of both scientific and industry know-how, demonstrating a clear understanding of both the innovation underlying the start-up and the business models available and needed to lead it to success. In short, the ideal investor not only provides the financial resources, he also offers support and advice.

Importantly, the prospective investment partner should fit the vision and philosophy of the start-up and be willing to grow with the company. He should be trustworthy, someone whose words can be taken at face value and who has a genuine interest in the development project in question.

3.4.2
The Venture Capitalist's Perspective

There are two major aspects venture capitalists evaluate before they will decide to invest in a biotech start-up. These are a sound scientific rationale and an excellent management team. Thus, investors are looking for a convincing, cutting-edge project with an innovative drug target, ideally one whose feasibility has already been demonstrated in proof-of-concept studies and that is based on sound science, perhaps even with a new technology platform. Ideally, the start-up also already has an experienced management team characterized by entrepreneurial thinking, making available a business plan that is convincing in terms of timelines, financing needs, and risk assessments. Only if these two factors coincide does the company stand a reasonable chance of surviving its founding years. These observations very likely derive from the fact that experienced investors have seen companies fail not only for scientific reasons but also as a result of wrong management decisions and conflicts among the management team.

Other factors investors mentioned were a unique selling proposition (USP), the credibility of their prospective partner, and solid intellectual property protection, all of which relate back to the scientific rationale of the project and the company's management team. Particularly the uniqueness of the company, i.e., its focus on an unmet medical need with a real competitive advantage, and an intellectual property (IP) situation securing the company the rights to its technological developments will be responsible for sustained and long-term market opportunities for the biotech start-up.

3.4.3
The Perspective of Other Investors

For the group of government funds and other experts or investors, the results were similar to those obtained in the group of venture capitalists, with a good management team, USP, and sound science ranking highest. The focus on the uniqueness of the business model suggests that, for this group, the 'market' factor is essential for the sustained success of the start-up to clearly differentiate its products from those of its competitors.

Overall, all three groups were found to rely most heavily on factors such as experience and the scientific background of their potential partner. Thus, the entrepreneur will look for an investor who understands the scientific rationale behind his invention and who has experience in financing start-up companies. The venture capitalist will be most interested in investment targets whose science and management team combine to provide a solid basis for future growth.

3.5
Financing Models

The second part of the questionnaire was dedicated to the importance of the different financing models available to European biotech start-ups. The goal was to identify the differences between these financing forms and to evaluate their individual strengths and possible drawbacks. The questionnaire listed the available financing models in the following order:

- Venture capital
- Corporate venture capital
- Private equity
- Strategic alliances with pharmaceutical companies
- Mezzanine capital
- Debt financing (banks)

- Government grants
- EU grants
- Non-profit organizations and foundations
- Angel investors
- Others.

All three interview groups were asked what they thought were the three most important financing models throughout the development of a European biotech start-up. Also, all interview partners were asked what they thought the major strengths and weaknesses of their top three approaches were and what they considered the benefits and drawbacks of each.

3.5.1
Most Important Financing Models

3.5.1.1   The Entrepreneur's Perspective

Top Three Financing Models
The relative importance of the individual financing models from the entrepreneur's perspective is given in Table 3.4, the top three financing models being venture capital, strategic alliances, and government grants.

Ranking fourth were 'other' financing models, with interview partners especially referring to a particularity among financing options called 'atypical silent partnership.' A silent partnership ('stille Gesellschaft') is a mezzanine financing instrument that has elements of both equity and loans. A company, provided with capital for a longer period, has to pay interest and in some cases profit-related compensation, making a silent partnership similar to a long-term bank loan. Yet, unlike a loan, a silent partnership is often subject to a subordination clause, giving the capital provided by the silent partner the status of equity despite its loan character. The silent partner provides capital but remains in the background and does not actively

Table 3.4 Importance of financing models from the perspective of the entrepreneur.

Financing model	*n*
Venture capital	10
Strategic alliances	7
Government grants	6
Others	5
Private equity	4
EU grants	4
Corporate venture capital	1
Mezzanine capital	1
Angel investors	1
Debt financing (banks)	0
Non-profit organizations	0

participate in or influence the management of operations. It can be contractually agreed upon that the silent partner does not participate in any losses incurred. Thus, the investor can assert his claims in the event of default, but the repayment is subordinated and all other lenders are re-paid first. In an atypical silent partnership, the investor participates not only in the profits or losses as well as in the value increase of the company. The atypical silent partnership is subject to a subordination clause. Participation in management and risk is an inherent part of the atypical silent partnership. Typical and atypical silent partnerships are not well known outside German-speaking countries.

For investors, the advantage of this model is that it allows them to harness tax benefits through tax-free investment of a part of their profits in high-risk areas, such as the biotech industry. For entrepreneurs, the obvious benefit is early and easy access to capital without any interference from the investor's side. In some European countries, such as Germany and Austria, it is not uncommon for start-ups to profit from this mode of financing in very early stages of their development.

Strengths and Drawbacks of the Top Three Financing Models

Venture capital   One of the main benefits of venture capital as indicated by the group of entrepreneurs is that it offers the only possibility to raise significant amounts of money under very risky circumstances. Thus, venture capital is the only way to raise €50–100 million within a period of 3–5 years. This gives entrepreneurs time to get the fledgling company on a sound footing while being able to look for additional capital sources. Moreover, venture capital helps entrepreneurs establish networks in the industry and build up expertise by cooperating with time-tested professionals. For entrepreneurs, receiving venture capital is perceived as a clear benchmark, and there is general agreement that it will be very difficult for a biotech start-up to sustain its long development times successfully without venture capital. Moreover, a positive company outlook combined with the interest of a venture capitalist will cause other venture capitalists to follow suit.

The major drawbacks as pointed out by the entrepreneurs are that venture capital creates a rather rigid business environment, limiting the decision-making capacity of the entrepreneur, who in some cases may even lose control of the company. Also, both partners enter into a long-term pact without knowing each other. Venture capitalists may pursue a short-term, exit-oriented approach. With investors not necessarily sharing the same interests and backgrounds as the company founders, they may also wish to interfere with the scientific questions at hand. In a nutshell, venture capital dilutes the company capital – and potentially also the decision-making power of the entrepreneur. Some entrepreneurs noted that they saw a tendency for investors to request increasing amounts of preclinical or clinical data before they decided to step in. Also, venture capitalists may be wary of investing in highly risky development phases. Finally, even though it is acknowledged that venture capitalists take on significant risk, they also gain high returns for doing so.

Strategic alliances   Strategic alliances with pharmaceutical companies allow entrepreneurs to cooperate with highly experienced and focused partners who are part of an existing network. Also, deals with well established companies guarantee a high degree of visibility. For example, a press release by a major pharmaceutical player is likely to have more weight and a wider reach than that of a small and still unknown start-up. Young companies will profit from the marketing and sales infrastructure already in place in large pharmaceutical companies. Thus, considering that the premarketing phase starts once Phase III trials are initiated, larger pharmaceutical companies may give start-ups access to key opinion leaders who they can involve in their clinical trial programs. All of these factors taken together may ultimately translate into shorter times to market.

The disadvantages include the sometimes dominating role pharmaceutical companies take on when cooperating with start-ups. Also, young companies run the risk of losing their appeal for other pharmaceutical companies once a deal with one partner has been closed. Finally, alliances may also result in competitive situations, e.g., in case of Phase III products that look highly promising: seeing that the clinical trial program is successful and the drug candidate holds great promise, venture capital firms may be even more willing to invest – whereas pharmaceutical companies may want to buy.

Government grants   One benefit of government grants is that they provide a very early source of start-up capital and are usually very well organized. Some interviewees stated that only public funds were capable of leading newly established companies through the early research development phase, because it generally takes 4–6 years, or at least US$10 million, to reach the preclinical data generation point.

From the perspective of the entrepreneur, one disadvantage of public monies is that government funding organizations usually have a less business-oriented approach and are therefore less capable of providing strong guidance in business-related questions. Government funds will beneficially support a company founder only if his strategy has already been clearly defined beforehand. Moreover, the combined grant and loan models applied by many European government funding organizations are not always regarded as beneficial compared to a pure grant model – at one point, any loan will have to be paid back, potentially leaving the entrepreneur without adequate funding at a time when he may need it most.

3.5.1.2   The Venture Capitalist's Perspective

Top Three Financing Models
The importance of financing models from the venture capitalist's perspective is given in Table 3.5. Thus, as for entrepreneurs, the three most important financing models are venture capital, strategic alliances, and government grants.

Ranking fourth was the group of business angels, who step in rather early in the financing process, particularly in the Anglo-American culture.

Table 3.5 Importance of financing models from the perspective of the venture capitalist.

Financing model	n
Venture capital	15
Strategic alliances	9
Government grants	7
Angel investors	4
Corporate venture capital	2
EU grants	2
Private equity	1
Mezzanine capital	1
Debt financing (banks)	0
Non-profit organizations	0
Others	0

Strengths and Drawbacks of the Top Three Financing Models

Venture capital   The fact that a company has been considered worthy of support by a venture capitalist firm is generally considered the result of a thorough and positive selection process that separates the wheat from the chaff. Venture capital firms are highly target- and return-oriented, pushing the limits of a company and offering added value in matters of management and business administration. With the exception of industrial contract manufacturing or the sale of technologies, biotech start-ups are unlikely to be successful without venture capital. Another advantage of venture capital is that company founders do not need any personal securities. Venture capital is true 'private equity' from professional partners with good networks and profound experience from previous projects, or a 'corporate memory,' enabling them to give meaningful strategic recommendations. Venture capital is available to companies involved in the high-risk area of the life sciences that are in too early and risky a phase for pharmaceutical companies to get involved. Venture capital is a highly recognized form of investment, also with major pharmaceutical companies, and a sign of quality for the supported company. Venture capital firms are flexible and open towards changes to the business plan should the circumstances so require, and will in most cases not attempt to exert any governance.

In terms of drawbacks, venture capitalists confirmed that funds were hard to come by, with fierce competition among applicants and only a small fraction of start-ups actually receiving venture capital. For the entrepreneur, venture capital is expensive money because investors will reserve the right to exert considerable controlling influence in the company. As such, equity participation dilutes the entrepreneur, who may have to give up certain of his rights. It is therefore essential for the founder to be given adequate incentives and sufficient shares to stay in the company. Conversely, if the cooperation does not appear to be working, investors will be adamant in taking the decisions it takes to lead the company to success, even if this means exchanging members of management who had been instrumental in founding the company. Also, investors can be rather inpatient, wanting to see fast successes and quick returns. Last but not least, any slowdown in the economy and downturns in the financial markets have direct repercussions on the venture capital industry and the availability of funds, making it difficult for investors to work in an anti-cyclic manner and potentially leaving young businesses without the support they need.

Strategic alliances   In the eyes of the venture capitalist, strategic alliances have the advantage of enabling intensive scientific cooperation with regard to validating the newly developed technology or product. Alliances provide non-dilutive capital and access to existing marketing and distribution channels, being a fantastic instrument for commercialization of the new product. Also, it is a positive sign for a young start-up to be taken seriously by a pharmaceutical company, and this may help attract additional investors.

On the downside, partnering with a big player too early may be tantamount to giving away much of the company's assets at a significant discount. Well established pharmaceutical companies, having to follow their internal 'chain of command,' may be slower to react than other types of investors. Pharmaceutical companies will invariably want to pursue their own interests, e.g., if a newly developed and promising molecule is at stake, being less interested in the sustainable growth and wellbeing of the start-up. As a result, the entrepreneur may have abandon much – or all – of his freedom and independence.

Government grants   In the venture capitalist's opinion, government grants are easy money that is inexpensive and non-dilutive for the entrepreneur. Government funds are usually quite entrepreneur-friendly and unbureaucratic. They dispose of significant fund sizes, usually without applying much due diligence.

On the debit side, government funds may wish to exert their control and influence on the management team. Even though public grants provide significant amounts of start-up capital, they often fail to offer business-oriented know-how. If debt financing is involved, the entrepreneur may be faced with the challenge of having to pay back the loan at a time when the required capital is unavailable.

3.5.1.3   The Perspective of Other Investors

Top Three Financing Models
The importance of financing models from the other investors' perspective are given in Table 3.6. As in the other two groups of interviewees, the three most important financing models are venture capital, government grants, and strategic alliances.

Ranking fourth in this group were EU grants, which are seen as an important new additional financing means. Other forms were found to play a minor role only.

Strengths and Drawbacks of the Top Three Financing Models

Venture capital   Venture capital in the group of other investors is also regarded as 'smart money,' early risk capital with added value through networks, experience, and professional support which pushes technology achievements towards an exit and

Table 3.6 Importance of financing models from the perspective of other investors.

Financing model	*n*
Venture capital	12
Government grants	12
Strategic alliances	6
EU grants	4
Corporate venture capital	1
Private equity	1
Mezzanine capital	1
Angel investors	1
Others	1
Debt financing (banks)	0
Non-profit organizations	0

makes the company fit for an IPO. This group of interviewees, too, stated that it would be extremely difficult to cover the massive capital needs without venture capital – particularly in the field of biotechnology.

The disadvantages include the short- to medium-term orientation of venture capital. Also, most venture capitalists step in rather late in the development process, i.e., at a time when risk is shrinking, making it difficult for entrepreneurs to finance their capital-intensive research after the seed financing stage. Being exit-oriented, with investors wanting to see their money return adequate benefits, venture capital binds the entrepreneur to deadlines that do not necessarily fit into the company cycle. Also, competencies are shared between the entrepreneur and one or more investors, potentially leading to conflicts between these different interest groups.

Government grants   The advantages of government grants are that they provide early money for investment into a business idea without diluting the capital, a concept sometimes referred to as 'soft money.' They are an important source of seed and early-stage financing, i.e., at a time when other investors are too risk-averse. Also, having received government support is a sign of quality for a start-up company, potentially motivating other investors to follow suit. Some Austrian interview partners stated that Austrian government funds were fairly well equipped compared with some of their European counterparts in other countries.

The disadvantages of government grants are the time it sometimes takes to get access to these public funds, especially in case of EU grants, and that preparing submissions for grants requires significant investments, which are sometimes not possible without venture capital. Also, the group of other investors stated that government funds were rather selective. For example, in Austria only 10–15 start-ups are supported by governments grants every year, some six of which come from the biotech area. In addition, some interview partners stated that public money always needed some co-financing in the form of venture capital to guarantee the survival of the start-up in the transition from seed to the early-stage financing.

Strategic alliances   The benefits of strategic alliances are that they are a sign of quality for the young company, increasing its significance in the eyes of both experts, investors, and the public and providing access to highly developed distribution channels. If a well established company partners with a start-up, the proof of concept will most likely already be available.

On the down side, young companies give up much of their independence when entering a deal with one of their big brothers, binding them to strict timelines and milestones. Because pharmaceutical companies are often not willing to take too high a risk, such alliances are mostly limited to later-stage development, often when clinical trials are already in Phase III. Also, by entering a strategic alliance, the perception in the eyes of other pharmaceutical companies is that the start-up has already been sold and its exit is at hand.

3.5.1.4   Conclusion

In all three groups of interviewees, the three most important financing forms for European biotech companies were venture capital, government grants, and strategic alliances. In some countries, such as Austria and Germany, government grant and loan arrangements play an important part in the seed financing of biotech companies. In other countries, such as Switzerland and the UK, there are hardly any public funds available, but there are significant numbers of business angels and early-stage venture capitalists taking over the part government funds play in other countries. Therefore, the optimal mix of financing forms appears to be country-specific.

Two of the entrepreneurs interviewed reported to have successfully relied on private equity as the major source of financing, whereas two others had had the opportunity to partner with big pharmaceutical companies early on. Entrepreneurs appear to be aware that, if they succeed in establishing a financing round with an important partner, additional capital investors are more likely to follow. Yet, they also know that the down side of an early strategic alliance with 'big pharma' is that this might deter additional investors or pharmaceutical partners. Government funds are regarded as an easy-to-receive financing tool, or, as one of the interview partners put it, it provides entrepreneurs with "some money to play with."

Most venture capitalists considered venture capital the single most important financing source for biotech companies. It is regarded as 'smart money,' securing the entrepreneur not only capital but also experienced partners and strong networks. However, venture capitalists also conceded that this financing form had the major disadvantage of diluting the entrepreneur. Therefore, venture capitalists are well aware that company founders need to be 'incentivized' to remain focused and motivated.

The group of other investors also considered venture capital the most important single financing form, but stated that its major drawbacks were the short-term (exit) orientation of many venture capitalists as well as their late entry, leaving many companies after the seed financing phase unserved.

Apart from these major financing models, the picture becomes more diverse and complex for the remaining financing forms. For entrepreneurs in German-speaking countries, the particularity of the atypical silent partnership is an essential early-stage financing contribution for many start-ups. For many entrepreneurs, private equity and

EU grants are also important. This shows that company founders are quite innovative when it comes to financing their capital-intensive R&D research. Almost every interview partner in the entrepreneur's group had taken an individual approach to financing their respective companies, looking not only which money was most easily available, but also which option produced the highest strategic and long-term value for the company. From the venture capitalist perspective, the fourth most important financing form, whose importance will continue to increase, are business angels, who, at least in the Anglo-American culture, step in rather early and frequently. They support the entrepreneur with their knowledge, experience, and network in a very early phase, at a time when it is still too risky for the venture capitalist to step in. Not surprisingly, other investors consider EU grants a major new source of additional financing, even though they acknowledge that the process from submission to approval takes quite some time and requires a great amount of administrative work and financial resources. All other financing forms, such as corporate venture capital, mezzanine capital, and debt financing, were considered less important by all three groups.

Overall, the funding of biotech companies therefore rests on the three major pillars addressed by all interview partners, complemented by a number of additional sources used and combined differently by different companies.

3.5.2
New and Creative Financing Approaches

The questionnaire section on the different financing models also asked whether there were other, more creative, financing approaches available to biotech start-ups. Only half of the interview partners ($n = 23$) responded to this question, with answers fairly homogeneous between groups and including:

- Project financing
- Research collaboration alliances
- Contract research
- High net worth individuals
- New stock markets for small and medium-sized companies
- Combinations of venture capital and angel investment
- Combinations of bio-incubator and corporate venture capital.

Thus, the biotech financing world appears to be much more colorful and varied than originally expected against the backdrop of the general financing matrix (Figure 2.2).

3.5.2.1   Project Financing
Project financing, also referred to as the 'virtual biotech model' (Birner, 2007), is based on the belief that, at least in theory, any element in the innovation chain of a biotech or medical device firm can be outsourced. In reality, this model presents both investors and the management team with considerable challenges. A 'virtual company' is supposed to outsource every major component of drug discovery and development to a third-party contractor. The most obvious advantages of the virtual biotech model derive

from its less rigid cost structure and major flexibility advantages over traditional business models. Start-ups are sometimes confronted with spiraling costs and increasing complexity of preclinical and clinical trials. The virtual model allows management to tap a vast network of clinical talent – without actually employing them. Therefore, the virtual approach gives start-ups the flexibility to quickly adjust burn rates up and down over time by creating a highly flexible company capable of expanding or narrowing the scope of product development within short notice. The most obvious disadvantage is the loss of control accompanied with this model. Placing key tasks in the hands of contractors may lead to increased risk and uncertainty. Also, some believe that virtual companies can skimp on management expertise. To the contrary, virtuals require even more management guidance than traditional start-ups, with particular emphasis on expert project management skills. By applying the virtual model, one can expect log-order reductions in early capital requirements while maximizing efficiency. This, in turn, can result in enlightened product development and reduced timelines. Investors and companies looking to acquire promising start-ups should therefore pay closer attention to virtual biotechs, not least because these companies lower the risk associated with developing promising biotech products. Reduced risk means increased upside potential for all stakeholders (Broderson, 2005).

3.5.2.2 Research Collaboration Alliances and Contract Research

Research alliances between research-oriented companies focusing on complementary areas and contract research on behalf of corporate clients, e.g., for developing a specific platform technology or producing industrial enzymes, are of significant worth to start-ups and had already successfully been used by some of the entrepreneurs interviewed. Such strategies will prioritize cost-intensive and arduous in-house development while at the same time generating revenue by providing services that offer comparatively lower margins but afford steady profits.

3.5.2.3 High Net Worth Individuals

High net worth individuals are individuals or families with investable assets in excess of US$1.0 million. One famous example is the German industrialist and one of the co-founders of SAP, Dietmar Hopp, whose fortune is estimated to amount to US$1.2 billion and who is considered an important financier of the German biotechnology landscape. Another example of a very active biotech investor is Andreas Strüngmann in Germany. In 2005, he and his brother sold Hexal as well as a 68% stake in US Eon Labs to Novartis/Sandoz for a total of US$7.4 billion. According to Forbes (www.forbes.com), he was the world's 214th richest man in 2007, with a net worth of approximately US$4.0 billion.

3.5.2.4 New Stock Markets for Small and Medium-sized Companies

The advent of new stock markets, such as the French Nouveau Marché, the Italian Nuovo Mercat, the UK Alternative Investment Market (AIM), and Nasdaq Europe, was among the most important reforms of stock exchanges in Europe in the 1990s. These new stock markets aim at attracting innovative early-stage and high-growth companies that would not have been viable candidates to public equity financing on the main markets of European stock exchanges (Giudici and Roosenboom, 2004). However, Europe's news stock markets met with only limited success. Stock prices plummeted after the stock market bubble burst, and new markets came to suffer from poor liquidity. One particularly negative example is certainly the deep fall of the 'Neuer Markt' in Frankfurt, opened in March 1997. In its six years of operation, the Neuer Markt had treated unwary investors to more of a roller coaster ride than they had expected, and, despite intending to set higher standards of accountability and openness, it was frought with numerous scandals. Neuer Markt's operator, Germany's main stock exchange Deutsche Börse AG, formally closed the exchange at the end of 2003 and replaced it with two broad segments in the main exchange, which Neuer Markt companies are being encouraged to join now. The deep fall of the Neuer Markt caused little surprise. The market's steep plunge has been a traumatic experience for tech investors, some of whom lost billions of euros (Blau, 2002).

3.5.2.5 Combinations of Different Types of Investors

Other creative financing approaches mentioned by the interviewees were mixes of different types of investors working together on one company project, e.g., a venture capitalist and a business angel or a bio-incubator and a corporate venture capitalist. Bio-business incubation is a holistic support process helping fledgling biotech companies succeed commercially by providing entrepreneurs with a wide range of targeted resources and services, from financing and consulting services to technical equipment and lab and office space. Thus, there does appear to be a tendency for individual investors to share competencies and capital costs to finance the expensive R&D efforts of companies they believe in. This tendency is reflected in different investor types moving out of their usual investment comfort zones and deciding for earlier participation in companies with a high-growth potential through adequate risk and cost sharing.

3.5.3
Differences Between Early- and Late-stage Financing

Two additional, more specifically financial, questions were posed only to the groups of venture capitalists and other investors, but not to entrepreneurs. The interview partners were asked to indicate whether and in what ways the financing in early- versus late-stage biotech companies differed. The results are summarized in Table 3.7.

Thus, there are differences between companies in different development stages, and these fall into one of three areas, i.e., people, the market potential, and the technology. One of the key elements contributing to business success is the quality of the management team. Management decides which human resources are needed and how they are used. A highly motivated team with complementary skills is an essential prerequisite for the success of a company, particularly one that has only just been founded. This is also reflected in one of the answers given by a venture capitalist who stated that in the early stages, 90% are invested in people and only 10% in assets, whereas in later stages, 50% are invested in people and 50% in assets.

Table 3.7 Differences between early- and late-stage financing.

	Early-stage financing	Late-stage financing
Probability of raising capital	Generally lower	Generally higher
Capital	More expensive for the entrepreneur	Less expensive for the entrepreneur
Risk-return relationship	High	Low
Size of investment	Series A: €1.5–4.0 million	Series B: €5–10 million Series C: > €10–15 million
Type of financing model and investor	Friends, family, and fools, government funds, incubators, angel investors	Venture capitalists (often organized in syndicates), strategic alliances with pharmaceutical companies, corporate ventures, mezzanine capital before IPO
Investment target	Mainly in people, less importantly in assets	50% in people, 50% in assets
Team size of start-up	Very small, research-oriented	Larger, more business-oriented
Technology versus product	Only technologies available, but no product	Technology and product available
Proof of concept	Not yet available	Already available
Valuation	Generally lower	Generally higher

The proper assessment of the prospective market is another important distinguishing factor. The ultimate proof of concept of any commercial undertaking is a paying customer. Thus, every company must carve out for itself a place in the market environment where it can reach its goals and satisfy its customers, while at the same time continuously following, analyzing, and responding to the market dynamics.

The most important assets of a biotech company are its science and technology. Therefore, any emerging company needs a unique selling offering to either develop new markets or challenge existing competitors. Any sound assessment of a company therefore also requires the evaluation of its technologic and scientific foundations.

Every company passes through distinct development phases, or life cycle stages, with distinct challenges and requirements. Thus, the assessment factors must always be analyzed in the light of the development phase a company is currently in. Different stages call for different management skills, different market environments, different product components, and different technologic or scientific foundations (Frei, 2006).

3.5.4
Differences Between Early- and Late-stage Valuation Techniques

The follow-up question asked for a more detailed description of the differences in valuation techniques between early- and late-stage development. Valuating a company is always a challenging task, but this holds particularly true for early-stage biotech companies. 'Pre-money' valuation, which takes place before a company is financed, dictates how equity is split between an entrepreneur and the investor(s). Entrepreneurs who do not fully understand the need for proper valuation at this stage might not only find themselves at a disadvantage in negotiations with investors, they may also later on have no opportunity to correct a suboptimal pre-money valuation. One prerequisite for a company to be valuated is that this company must be a 'going concern,' i.e., a business that has a future perspective, functions without the threat of liquidation, and generates ongoing cash flow. Thus, the value of a start-up lies in its future potential rather than in its present state. However, because the future of a company depends on the strategic decisions taken today, these are important factors to consider when valuating a company. During the valuation process, assumptions about the future market perspectives, the major elements of the product, the scientific and technologic foundations, and the ability of the management need to be assessed (Frei, 2006).

Unlike the highly standardized process of financial accounting, there are no generally accepted rules or standards for valuating a company. However, there are a number of recognized methods, and investors generally use a combination of these, depending on the type of the business they invest in and on how far developed it is. It is difficult, for example, to apply conventional methods of valuation in the case of a start-up company that consists of intangible assets only. These start-up businesses have risky turnover and profit projections, and can hardly be compared to existing, well established companies (European Venture Capital Association, 2007a). In principle, valuation methods can be categorized into methods based on discounted cash flows, comparable methods, the venture capital method, real options, and the Monte Carlo method (Frei, 2006).

3.5.4.1 Methods Based on Discounted Cash Flows

The most common approaches to primary valuation in the corporate finance literature are generically referred to as the discounted cash flow (DCF) methods, whereby a company is valued at the present value of the future cash flows it will be able to generate. This method is also referred to as primary valuation, because it is based on such fundamental information as projected future free cash flow (FCF) and costs of capital. This method can be used when positive cash flows are able to be estimated with a reasonable degree of certainty and when the discount rate can easily be approximated. This rate is based on the rate of a risk-free investment, such as a state bond, to which a premium is added. The level of the premium depends on the estimated risk level. This discount rate is then applied to the cash flow over the investment period, and the sum of DCFs finally adds up to the current value of the company (Table 3.8). These methods are conceptually robust but can prove difficult to implement in high-uncertainty environments, such as those early-stage biotech companies find themselves in. Typical problems derive from the highly uncertain and distant positive cash flows, a business model based on many assumptions, and a difficult risk profile. The easiest approach to determine the most appropriate discount rate in a DCF calculation is one that uses the stage of development of the company, which can be determined by the drug development phase of products in the pipeline as proxies for the risk (Frei and Leleux, 2004).

Table 3.8 Discount rates to be used depending on stage of product development (Frei and Leleux, 2004).

Company stage	Discount rate (%)[a]	Drug development stage
Seed stage	70–100	Generating leads
Start-up stage	50–70	Optimizing leads/preclinical
First stage	40–60	Phase I clinical trial
Second stage	35–50	Phase II clinical trial
Later stage	25–40	Phase III clinical trial

[a]The discount rate presented incorporates the overall 'risk profile' of the company investigated, a profile driven jointly by technological, market, and business (i.e., management and organizational) risks (Frei and Leleux, 2004).

3.5.4.2 Comparable Methods

This valuation technique refers to all methods based on comparing one company to other – quoted or unquoted – companies in the same business sector in terms of factors such as profits, cash flow, net worth, or turnover. The comparable method is also referred to as a 'secondary' valuation method because it uses the market value of comparable companies or transactions as point of reference. Secondary valuation assumes that these comparable companies have been properly valued and can therefore serve as benchmarks when assessing a company (Frei, 2006).

The most frequently used comparative method is based on the price–earnings (P/E) ratio, also referred to as 'price multiple' or 'earnings multiple.' This is a valuation ratio of a company's current share price compared to its per-share earnings. The higher the P/E ratio, the higher the earnings growth expected by investors. For the method to generate reliable results, it is important to compare the P/E ratios of companies in the same industry. The underlying assumption is that markets have correctly assessed future values. Other comparative criteria, such as the price/EBIDTA ratio or the price/net worth ratio, are sometimes used in the same way. The valuation of companies with multiples is relatively straightforward and intuitively clear. The ease of use and the fact that they reflect the market mood makes multiples a very attractive method for valuating start-up companies. However, the simplicity may result in inconsistencies and vulnerability to manipulations (Frei, 2006).

Practitioners sometimes argue that DCF is much more complicated than the multiple method. However, this is not altogether true. With a DCF valuation, explicit assumptions are made, whereas with the comparable approach, implicit assumptions are taken and are often not stated. In the end, the value of a company depends on growth, ability to earn cash, and risk. This is also how market values are computed. In using comparable valuation, investors assume that the company to be valued is based on the same fundamentals as the comparables (Frei, 2006).

3.5.4.3 Venture Capital Method

The valuation of companies with negative cash flows and earnings can be quite difficult, as shown earlier. The venture capital method is a rather simple approach, trying to overcome some of the negative aspects of DCF valuation. It is a widely used technique by venture capitalists to value high-growth companies. In a first step, the exit value at the time of the company's IPO is calculated. This exit value is then used to calculate a net present value (NPV). Whereas a DCF valuation takes the view from the company side in looking at the possibilities to generate free cash flow to infinity, the venture capital valuation approach utilizes the view of the investor when calculating the value of the company, with venture capital investors generally looking for an exit within a defined time frame. Not surprisingly, this is why this approach is called the venture capital method (Frei, 2006).

3.5.4.4 Real Options
Real options capture the value of managerial flexibility to adapt decisions in response to unexpected market developments. Companies create shareholder value by identifying, managing, and exercising real options associated with their investment portfolio. The real options method applies financial options theory to quantify the value of management flexibility in a world of uncertainty. If used as a conceptual tool, it allows management to characterize and communicate the strategic value of an investment project. According to Amram and Kulatilaka (1999), the following types of option pricing valuation approaches can be distinguished: If a company can wait to invest, it can thus create value, because it gives the investor the opportunity to gather more information as time goes by and risk declines. If an investor, with an initial investment, can secure the possibility to access additional value with a follow-on investment, the term 'growth option' is used. The flexibility option provides value because it allows a company to make changes based on recent developments. An exit option provides its holder with value because a project can be abandoned if certain events take place (Amram and Kulatilaka, 1999). Traditional methods, such as NPV, fail to accurately capture the economic value of investments in an environment of widespread uncertainty and rapid change. Real options enable corporate decision makers to leverage uncertainty and limit downside risk.

3.5.4.5 Monte Carlo Method
The Monte Carlo method is based on the use of what are called random numbers, which can be generated by a roulette wheel. It is perfectly suited to model uncertainty, especially when there are many sources of uncertainty and the problem is complex. The basic idea is to generate the distribution of possible outcomes by simulation. In classic Monte Carlo, the random numbers are independent and identically distributed. Today, the Monte Carlo method is widely used in risk management applications. It is a powerful, flexible method and generally easy to implement. As a drawback, it is sometimes associated with computational burden for large-scale problems and might then require a lot of work (Boyle and Boyle, 2001).

The relative importance of valuation techniques in early- and late-stage financing of biotech start-ups as seen by the interview partners is given in Tables 3.9 and 3.10. Multiple answers were allowed. In a next step, the overall results on early- and late-stage financing were compared (Figure 3.1).

Table 3.9 Valuation techniques used in early-stage financing of biotech companies.

Early-stage valuation techniques	n
Comparables	9
Venture capital method	6
Evaluation of management and staff	3
Gut feeling	3
Real options	2
Discounted cash flow	1
Monte Carlo	1

Table 3.10 Valuation techniques used in late-stage financing of biotech companies.

Late-stage valuation techniques	n
Comparables	9
Venture capital method	7
Discounted cash flow	4
Real options	2
Monte Carlo	2

The valuation techniques considered acceptable in earlier development phases are rather diverse, ranging from comparables, the venture capital method, valuation of the management, and gut feeling to discounted cash flow, real options, and even Monte Carlo analysis. In later development stages, there is a shift to fewer and more sophisticated techniques, i.e., DCF, Monte Carlo simulations, or real options.

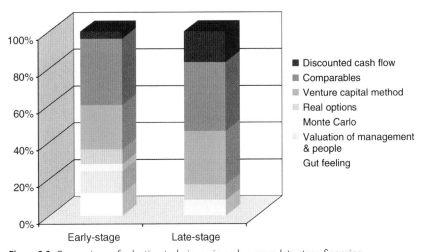

Figure 3.1 Comparison of valuation techniques in early- versus late-stage financing.

Valuating high-growth high-uncertainty companies is a challenge, particularly if they are combined with accounting losses. Even though valuating such companies is a challenge, particularly if they are combined with accounting losses, discounted cash flow remains a useful tool. Alternatives, such as the price-earnings multiples, generate extremely imprecise results (earnings are highly volatile), often cannot be used (when earnings are negative), and provide little insight about what drives the company's valuation. More important, these shorthand methods cannot account for the uniqueness of each company in a fast-changing environment. Another alternative, real options, is certainly promising, but current implementation techniques still require estimates of the long-term revenue growth rate, long-term volatility of revenue growth, and profit margins – the same requirements as for discounted cash flow (Koller et al., 2005).

Therefore, the ideal valuation method that is universally applicable does not exist. However, by combining different methods, it is possible to compare and review results and to make adjustments where needed. Moreover, applying various techniques will result in a more differentiated picture of the overall company value. Combining various methods was also mentioned as the most meaningful option by the interview partners. They were keenly aware of the difficulties involved in valuating start-ups, particularly those still in an early stage, and that a thorough company valuation requires high-level expertise in valuing risk and applying the available techniques – regardless of how they are combined.

3.6
Biotech Research Areas

Another section in each of the three questionnaires was dedicated to the different research areas in the biotech industry. Interview partners were asked which areas they found most promising, with multiple answers allowed. There were no major differences between the three groups, even though, quite understandably, entrepreneurs tended to favor their own research areas. Overall, interviewees did not have any specific interest in only one or two particular research areas. Some experts pointed out that all indications may be of interest, as long as the following caveats are considered.

Thus, in terms of market potential, the market size and the innovative potential of the drug candidate are important. Generally, any area with an unmet medical need can be an attractive investment target. Another essential aspect is the treatment duration, with indications requiring longer-term treatment being more attractive than short-term therapies. Here, cancer treatments are still high on the list of attractive drug candidates, and many experts are convinced that cancer will soon be a chronic disease rather than a fatal diagnosis. At the same time, interview partners stated that there was hardly a research area more risky and challenging than oncology. Finally, adequate intellectual property protection needs to be put in place, and the business model must be sales-driven.

Promising indications also include those that attend to upcoming or future healthcare trends and that will gain importance in the future, such as disease prevention (e.g., through vaccines against cancer), the prediction of disease, and personalized medicine (Burrill, 2007). One of the slogans of the future will be 'from sickness to wellness.' Also, it will be important to address the needs of an aging population and to develop means to foster health management in the elderly.

Particularly from the vantage point of the venture capitalist, the portfolio should always be as diversified and balanced as possible to spread and reduce risk across the different biotech companies he has invested in so as not to have to consider too many details per company, product, and market. Some interviewees seemed to express a higher level of interest in specific indications, such as immunology and oncology, and appeared less interested in areas such as neurology, cardiovascular disease, pain, and respiratory diseases, although trends in funding different indications usually follow a cyclical up and down behavior.

3.6.1
Vaccines

A research area that has come to enjoy renewed interest and that was mentioned by almost half of the interview partners was the preventive and therapeutic usage of vaccines. Especially treatments against incurable cancer remain an area of particular interest for vaccines, which are now coming into their own with new technologies (Ernst & Young, 2007). Another reason for this revival is that, in contrast to the past, the economies for vaccines are beginning to make sense, not least because of new pricing strategies positioning these products in high-end price categories. Therefore, more and more start-ups focus on the development of new vaccine compounds. In addition, an increasing number of pharmaceutical companies buy in vaccine pipelines. One example for this trend are the recent investments of the Novartis Venture Funds into two European firms developing vaccines, amounting to a total of €18.2 million (approx. US$22.8 million; Scrip, 2007a). Another example is the US$15.6 billion (approx. €12.5 billion) acquisition of MedImmune, a start-up company with a strong vaccine focus, by AstraZeneca, a company not previously involved in the development, sale, or marketing of vaccines (Scrip, 2007c). Despite this recent trend, the number of players in the vaccines industry is still limited, with all the particularities of an oligopoly market dominated by a small number of participants who are able to collectively exert control over supply and market prices. Some internationally recognized companies show highly promising financial figures, such as the Austrian vaccines company Intercell or the major pharmaceutical companies GlaxoSmithKline, Merck, and Sanofi–Aventis (Gillinger, 2007).

Regarding the possible differences in the business models between a classic biotech product, such as antibody development against cancer, and a vaccine, interviewees stated that business models should take into account possible differences in three areas, i.e., R&D, manufacturing, and market analysis. The following questions are therefore worthwhile asking:

Research and Development

- Are there any government-induced incentives to start research on a particular vaccine, e.g., a vaccine against HIV?

- Can a partnership with a government or non-profit organization be established? Examples of organizations include the NIH, the military, or the Bill and Melinda Gates foundation.
- Does the clinical trial program focus on prevention or cure?
- How many subjects are needed for clinical trials? For example, for childhood vaccines, large sample sizes will be required.

Manufacturing

- Should an early strategic partnership for manufacturing collaboration purposes be entered into?
- What are the production capabilities?
- What are the manufacturing costs?

Market Analysis

- How large is the number of potential buyers?
- How many possible competitors are there?
- What is the market potential? Is the vaccine effective against an infection or disease prevailing in only some parts of the (developing) world, e.g., West Nile virus infection, or will the vaccine be able to be sold on a global level, e.g., a hepatitis vaccine?
- What pricing strategy will be pursued?
- How likely is the product to receive recommendation or reimbursement in the countries where it will be marketed?
- How can vaccination be provided to and financed in developing countries?
- How will the vaccine be commercialized? Will a distribution partner be required? Will it be possible to conclude tender agreements with governments?

3.6.2
Biotech Drug Development or Medical Technology and Devices?

Many interview partners raised concerns that, compared to biotech drug development programs, research into medical technology or medical devices may be more beneficial due to more rapid and less risky product development and shorter times to market. However, they also acknowledged that the number of such companies was rather limited and that, here too, funding depended on the possibilities of earning an attractive return. The challenge in the area of diagnostics and medical devices is that margins are not always attractive enough for investors to step in. General examples of recent successful research initiatives mentioned included telemetrics, neurostimulation, and head-borne microscopes.

Some interview partners stated that investors may tend to favor diagnostics and medical devices because these were easier to understand than the complex molecular pathways and target-oriented small molecules interacting with defined cascade mechanisms in the human body. Particular interest was expressed in reliable diagnostics, e.g., for early identification of diseases, such as cancer, biomarkers, radioimmunodiagnostics, new imaging techniques, or theragnostics – individualized therapies indicating early on which patient will be eligible for which treatment.

3.7
Alliances and Funding in the Early Stage

3.7.1
Networking Between Venture Capitalists and Universities and Their Scientists

The first item in the questionnaire section dealing with alliances in the early stage of a start-up asked whether (strategic) alliances between venture capitalists and universities and their scientists were important and if so, how such networks could be established. The results were quite diverging between groups, as displayed in Figure 3.2.

The ideas of how such networks and alliances could be established were similar across all three groups:

- Establish early, continuous platforms and networking events between scientists, institutions, venture capitalists, government bodies, etc.
- Provide information on how to set up a business plan
- Organize business plan competitions
- Provide information on IP protection
- Enhance tech transfer offices at universities
- Establish commercial intermediaries, such as incubators and business angels, and mentor young entrepreneurs, e.g., through tech transfer offices
- Invite scientists to venture capital conferences, give incentives, establish personal contacts

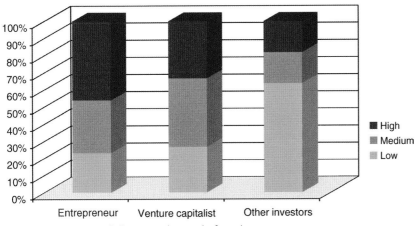

Figure 3.2 Importance of alliances and networks from the perspectives of the entrepreneur, the venture capitalist, and other investors.

- Teach students at university the do's and don'ts of commercializing an idea, hold practical training sessions, introduce the concept of venture capital to university students
- Attract talent from university for master or doctoral theses and provide job offers later
- Attract researchers from abroad
- Share resources between universities, e.g., laboratories
- Develop regional clusters to find entrepreneurs and give them access to networks
- Use successful role models
- Present success stories, e.g., at networking events between investors and universities
- Hire more investment managers with technological background.

3.7.1.1 The Entrepreneur's Perspective

From the entrepreneurial side, the desire to become part of such networks appears to be higher than from the venture capitalist side. For the entrepreneur, such partnerships increase the likelihood of getting earlier access to capital and profiting from networks for research activities, in turn increasing and enhancing their own R&D productivity and potentially leading to a company or university spin-off. Overall, the need of the entrepreneur for professional consulting is quite high, particularly with regard to putting together a business plan, protecting his intellectual property, or enabling him to access various networks. These needs must be addressed quickly and efficiently together with the entrepreneur to help him get his company up and running.

3.7.1.2 The Venture Capitalist's Perspective

For the venture capitalist, the scientist fulfils an important external consultant function, helping him evaluate which projects and companies may be of interest for investment activities. Many venture capitalists therefore have a pool of experts to consult with on new technologies or platforms. They generally do not see a need to put such collaborations on a broader basis, because they believe that scientists and future entrepreneurs will find their way to an investment company if they have an interesting R&D project at hand that may lend itself to being developed into a biotech start-up. The huge numbers of business plans venture capital companies receive every year for evaluation and possible funding bear ample proof of this.

3.7.1.3 The Perspective of Other Investors

From the point of view of other investors, early alliances should be initiated by incubators and preseed government funds rather than by venture capitalists. They claim that there is already a proven process in place and that venture capitalists are not usually prepared to establish contacts or networks with a scientist and to fund biotech start-ups in the preseed phase.

It would certainly be a worthwhile investment into the future to start the process of entrepreneurship for biotech companies at university to make students aware of what it means to found a company. Commercialization of a good idea could effectively be initiated at a university that offers support, coaching, and mentoring functions as well as institutions that have the knowledge and the experience to support the founding process, such as tech transfer offices, bio-incubators, or business angels. For such a concept to become a viable option in Europe, many universities will first have to open up to the outside world and to see the opportunities that are related to commercializing inventions and setting up a university spin-off.

3.7.2
The Role of Government Grants

One questionnaire item was related to the importance of government grants. For all interview groups, the positive effects of such grants outweighed any possible drawbacks. For the groups of entrepreneurs and other investors, there was common agreement that government grants are highly important for a biotech start-up. Only some venture capitalists viewed this funding source with reservation. The importance of government grants as rated by all three groups of interview partners is shown in Figure 3.3.

Also, the interview partners were asked what they considered the most important government support model. From the perspective of entrepreneurs and other investors, a mix of grants and loans as provided in many European countries was the most favored government support model. Only the venture capitalist was predominantly in favor of pure grants (Figure 3.4).

With regard to government funding, the answers produced rather controversial results between groups and depending on the country the interview partners lived and worked in. For the Swiss, government grants and loans are rare and play an inferior role. This is due to the already high density of biotech and pharmaceutical companies and university spin-offs newly founded every year and a higher frequency of business angels who step in quite early in the development process. Also, the last years have

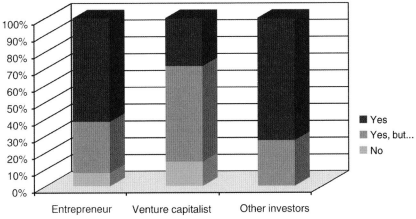

Figure 3.3 Importance of government grants from the perspectives of the entrepreneur, the venture capitalist, and other investors.

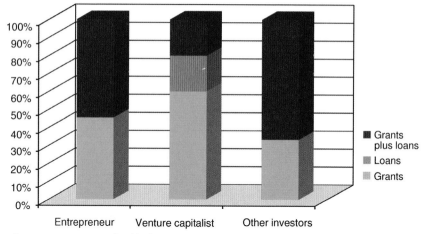

Figure 3.4 Importance of public fund models from the perspectives of the entrepreneur, the venture capitalist, and other investors.

seen numerous Swiss biotech start-ups write their stories of success, such as Actelion and Speedel, and these positive examples have further spread the knowledge of how to successfully start a biotech company.

In another highly advanced venture-capital-backed country, the UK, the situation resembles that seen in Switzerland. Thus, there are only few public funds available, but it is generally easier than in other European countries to find and cooperate with professionals knowing the biotech start-up business inside out and to profit from a larger number of private investors. In other countries, such as Austria and Germany, entrepreneurs heavily rely on government grant and loan models established by their governments to promote research, innovation, and entrepreneurship.

3.7.2.1 The Entrepreneur's Perspective

From the entrepreneur's perspective, government grants and loans are highly important as preseed and seed financing tools. However, they are aware that the best grant and loan funding would be of no benefit without a comprehensive business plan that defines the long-term company strategy. Today, government organizations are better prepared and equipped than in the past to support future entrepreneurs in determining the most suitable strategy for their companies, not only in terms of developing a business plan, but also in helping the founder translate this plan into reality and in accompanying and supporting the start-up to reach its milestones.

3.7.2.2 The Venture Capitalist's Perspective

Even though all interview groups agreed that the positive effects of government grants outweighed any possible drawbacks, there were also some critical remarks. Specifically, venture capitalists mentioned some caveats with regard to loans, stating that loans and atypical silent partnerships may have a negative impact on the balance sheet by diluting the company value. Others thought that government funding mechanisms with grants and loans should only be put in place after a venture capitalist has evaluated the company and decided to step in. Some other venture capitalists said they were skeptical with regard to the evaluation process of government organizations and believed that too many start-ups received preseed and seed financing, even though many of them were not going to be able to raise follow-on capital. Also, venture capitalists believe that a more stringent process on the part of public bodies in supporting start-up companies would be beneficial for the entire funding process. At the other end of the spectrum are venture capitalists who believe that, in general, government organizations should not offer public money to support the seed financing phase at all, because such 'donations' only offered short-term support without encouraging entrepreneurs to develop a true long-term commercial perspective on where the company should go in the future.

3.7.2.3 The Perspective of Other Investors

From the perspective of other investors, grants are usually regarded as important, particularly when they are invested in basic research. Most interviewees in this group considered the government funding mechanism an important tool and a sign of quality for the young company, particularly for follow-on investors. Of interest, some venture capitalists expressed the opposite opinion, namely that grants and loans are detrimental to the start-up development, because they treat the entrepreneur to some easily obtained money but do not motivate him to raise the value of this company.

3.7.3
The Role of Business Angels

In view of the prospering business angel situation in the US, a comparison to the situation in Europe was considered interesting, and one part of the questionnaire was therefore dedicated to the role of business angels. The proportions of interviewees having established contacts with business angels are presented in Figure 3.5.

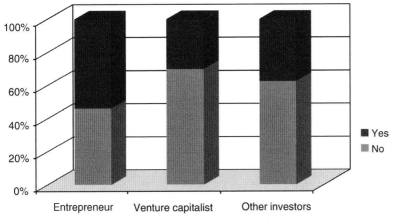

Figure 3.5 Partnerships and cooperations with business angels from the perspectives of the entrepreneur, the venture capitalist, and other investors.

Europe is slowly but continuously catching up with the US and its wide network of angel investors. Again, Switzerland and the UK are far ahead of other countries when it comes to angel investments. Business angels usually act at a local level, are heavily involved in the development of the companies they invest in, and do not generally invest in companies abroad.

In other European countries, both entrepreneurs and venture capitalists were quite optimistic with respect to angel investments, indicating that, even though these played only a moderately important role today, they could become more significant in the future. The most skeptical group in this context were the other experts, with two thirds stating that they considered the importance of business angels negligible (Figure 3.6).

3.7.3.1 The Entrepreneur's Perspective

For entrepreneurs, having an early-phase investor obviously is of the essence, and more than 50% of company founders interviewed reported already having established relationships with business angels (Figure 3.5).

One additional advantage entrepreneurs mentioned was again related to the fact that, if a company succeeds in getting a business angel to step in early in the development process, this is likely to attract additional investors.

3.7.3.2 The Venture Capitalist's Perspective

From the venture capitalist perspective, alliances with business angels are taken with a grain of salt. Many venture capitalists believe that once a second, bigger financing round needs to be established, the cooperation with a business angel may become difficult, because business angels are usually interested in keeping their influence in the company without, however, being able to bring in more capital. Moreover, according to some venture capitalists, there have been some negative experiences

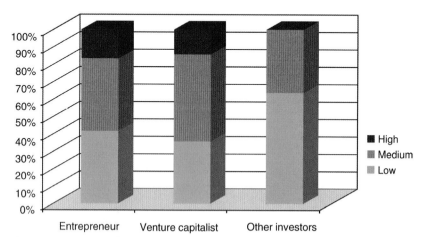

Figure 3.6 Importance of business angels in Europe from the perspectives of the entrepreneur, the venture capitalist, and other investors.

with European business angels, asking for 10% of the company in return for mere consultancy services without offering any financial contributions at all. Overall, the majority of venture capitalists consider business angels in biotech as moderately important due to the huge amounts of capital needed, the long development periods, and the high medical and technical expertise required to understand sophisticated, cutting-edge research projects.

3.7.3.3 The Perspective of Other Investors

Other investors also had a sense that the rare business angel expressing an interest in biotech was not always welcomed by venture capitalists – for all the same reasons already stated by venture capitalists themselves.

3.7.4
Preferred Deal Structure

The structure of the deal between the entrepreneur and venture capitalists determines the legal and financial framework of their cooperation. Equity-based compensation plans generally make up a significant amount of the rewards made available to the entrepreneur and his team. Usually, the investor will pose what is referred to as a 'vesting period.' This requires the stock option beneficiaries to stay in the company for a minimum amount of time to not cause the share price to falter. Once the vesting period has expired, the entrepreneur can profit from the preferential conditions of stock options. The contract between entrepreneur and venture capitalist will also contain a shareholders' agreement, specifying the rights and obligations of each party (e.g., the reciprocal right of veto on certain decisions, the transfer of preferential shares to venture capitalists, the policy of distributing dividends, the composition of the Board of Directors, and the timetable of business and financial reporting), the investment protocol (i.e., the price and number of shares), changes required to the statutes, warranty letters, employment contracts for the management team, and so forth (European Venture Capital Association, 2007a).

The venture capitalists and the group of other investors were asked what they considered the preferred deal structure. There was common agreement among almost all interview partners what the most favorable terms for the investor were, regardless of the development stage of the company. Thus, the two most popular arrangements with entrepreneurs were convertible preferred shares and a liquidation preference.

3.7.4.1 Convertible Preferred Shares

Convertible preferred shares are bond-like securities that are senior to common stock and can be converted to common stock. They enjoy certain preferential rights, such as an anti-dilution provision, giving venture capitalists the right of first refusal over any future issue of equity to outside investors. In such an agreement, the anti-dilution provision adjusts the number of shares (or the percentage of the company) held by the holders of the preferred shares upwards if the firm subsequently undertakes a financing round at a lower valuation than the one at which the preferred investors purchased the shares.

Venture capitalists often receive anti-dilution provisions on their preferred stock. In theory, such protection gives a venture capitalist claims on additional shares in the event of a down round, a financing round in which stock can be purchased at a lower valuation than that placed on the company by earlier investors. Such down rounds would otherwise cause dilution of ownership of existing investors. In practice, however, investors in struggling companies are usually forced to give up these rights to make a new round of investment possible (Lerner et al., 2005; Metrick, 2007).

3.7.4.2 Liquidation Preference

A liquidation preference determines the order in which different security holders are paid in the event of a liquidation. It ensures preference over common stock with respect to any dividends or payments in association with the liquidation of the company. By this means, the investor knows where he stands in the capital-structure hierarchy. In recent years, it has become popular for venture investors to insist on liquidation preferences in excess of their original investment. For example, a $2\times$ or $3\times$ liquidation preference means that the investor will be paid back double or triple their original investment before any of the other junior equity claims are paid off (Amis and Stevenson, 2001; Lerner et al., 2005; Metrick, 2007).

Overall, there appears to be common agreement among venture capitalists and other investors about what must be covered in a term sheet to maximize and optimally protect the investors' interests.

3.7.5
Structuring the Board of Directors

All interview partners were asked what they regarded as the most useful approach to structuring the Board of Directors.

3.7.5.1 The Entrepreneur's Perspective

For entrepreneurs, the mix of expertise was the most important attribute of the Board of Directors. Ranking second, third, and fourth were the small size of the board, its independence, and the willingness and ability to actively contribute to decision-making process (Table 3.11).

From the answers given by entrepreneurs it becomes evident that they attach major importance to diversity within the Board of Directors, bringing together personalities with a wide range of skills and experiences. At the same time, they prefer small boards to speed up the decision-making process.

3.7.5.2 The Venture Capitalist's Perspective

Venture capitalists considered independence the most important attribute of a Board of Directors (Table 3.12), followed by active contribution, mix of expertise, and small size.

Table 3.11 Most important criteria for structuring the Board of Directors from the entrepreneurs' perspective.

Criterion	Description
Mix of expertise	Mix of scientific, economics, and industry experts
	Mix of experts representing all important corporate functions to avoid 'blind spots'
	Rich diversity through top people
	Have some non-investor representatives
Small size	Keep the board small and inter-related to enable adequate networking
Independence	Independent experts who are not dominated by the interests of investors
Active contribution	People who know the tools of their trade
	People who take their time to contribute actively to the board
	People who have invested in the company and identify with the entrepreneur

Table 3.12 Most important criteria for structuring the Board of Directors from the venture capitalists' perspective.

Criterion	Description
Independence	Independent members making sure the company is operated in the best possible way
	Industry-experienced managers
	Free of political influence
Active contribution	People who contribute actively as opposed to individuals who are merely well known
	Proactive people who want to make things happen
	People who are influential in society
Mix of expertise	Different areas of know-how (e.g., biotech, IPO, regulatory, finance)
	Complementary team and skills
Small size	Keep the board small and flexible

Some interview partners stressed the idea that the board should be composed of experts and influential members of the society and that membership in the board should not be determined solely on the basis of the size of the investment. Also, some venture capitalists stated that the composition of the Board of Directors may have to be adapted to the different development stages of the company. For example, in the time leading up to an IPO, the venture capitalists may step back and leave the floor to experienced industry experts.

3.7.5.3 The Perspective of Other Investors
The group of other investors again considered the mix of expertise the most important criterion for structuring the Board of Directors, followed by independence and active contribution (Table 3.13).

Table 3.13 Most important criteria for structuring the Board of
Directors from the perspective of other investors.

Criterion	Description
Mix of expertise	Mix between economics/finance experts and pharmaceutical experts
	Members should be heterogeneous in terms of know-how and character
Independence	External experts who maintain their independence
Active contribution	People with a good culture of constructive controversy
	Discuss and agree on next financing rounds
	Ability to say things that might not please everybody

Interestingly, the size of the board was not an issue in this group. However, this group placed special emphasis on the ability to overcome conflicts and controversies as well as on the willingness of board members to be able to speak out openly, even when what they say may not always please everybody. Considering that even small boards combine personalities with many diverse experiences and educational backgrounds whose opinions may diverge considerably, a fair amount of problem-solving capacity of the members of the board is essential if stagnation within the company is to be avoided.

The ideal composition detailed by the group of other investors was similar to that given by entrepreneurs, with a mix of expertise most prominent, whereas the venture capital group ranked independence and the active contribution of each member highest. Overall, all three groups largely agreed on the major factors to consider when staffing the Board of Directors.

3.8
Alliances and Funding in the Late Stage

3.8.1
Networking with Corporate Venture Capitalists

The questionnaire section dealing with alliances in the late stage of a start-up asked how venture capitalists and other investors valued possible partnerships between venture capitalists and corporate venture capitalists further down the development path of a young company and how such networks and alliances could be established or strengthened. The group of entrepreneurs was asked a slightly different question, namely what a pharmaceutical company needed to offer an entrepreneur to motivate him to enter an alliance.

3.8.1.1   The Perspectives of the Venture Capitalist and Other Investors
The importance of strategic alliances between venture capitalists and corporate venture capitalists from the point of view of both groups of investors is displayed in Figure 3.7.

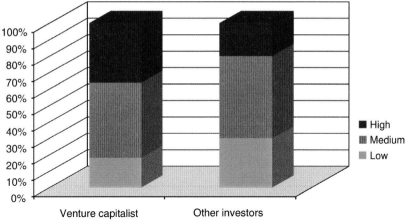

Figure 3.7 Importance of strategic alliances and networks between venture capitalists and corporate venture capitalists from the perspectives of venture capitalists and other investors.

When asked what could be done to enhance the development of such alliances, both groups agreed that partnering events, such as the senior-level conferences Bio-Europe Spring® or Bio-Europe, are best suited to establish contacts and guarantee maximum networking opportunities (EBD Group; www.ebdgroup.com).

With regard to later-stage (strategic) alliances between venture capitalists and corporate venture capitalists, such as the ones established by big pharmaceutical players such as Novartis, GlaxoSmithKline, Amgen, or Genentech, the answers obtained were very similar between both groups of investors.

In general, the alliance between venture capitalists and corporate venture companies is regarded as a beneficial option, as long as big pharmaceutical companies do not claim any special rights, such as the first right of refusal, the right of utilization, or the first option on the product. Rights issues could clearly become a conflict of interest, because not only is a corporate venture capitalist accountable to his investor, i.e., the mother company, he is also obliged to sell the company at the best achievable market price. Therefore, syndicating is desirable provided that no special conditions are to be granted to the corporate venture fund. Moreover, as is the case for cooperations with business angels, corporate venture capitalists will also have to accept that, if they cannot finance a second round, they will be diluted and outvoted.

As every other phase in the drug development and financing process, later-stage development is a 'people business', depending on individuals who are willing to build up mutual trust and rely on each other even in times of crisis. This is important, because the goals of the individual parties involved in getting a start-up up and running might not always concur. For example, whereas big pharmaceutical companies may be reluctant to share their knowledge for competitive reasons, the small biotech start-up may want to share their know-how to expand and strengthen their network.

The capital endowment of dedicated funds made available by pharmaceutical companies appears to depend on the cyclic behavior of the pharmaceutical industry.

As a result, such funds can rapidly dry up as the company's own cash flow decreases. Also, venture capitalists are generally wary of the 'technology scouts' working for some of the corporate venture capital funds that may be more interested in staying up to date with regard to novel and cutting-edge technologies being developed than in genuinely supporting a young enterprise.

The overall attitude of investors towards corporate venture capital is certainly positive. With well established pharmaceutical companies usually conducting thorough due diligence efforts, having a corporate venture capitalist step in is again considered a sign of quality for a young company. Yet, partners have to be carefully selected for the overall best interests of the start-up. In case of diverging interests, it may be better to forgo such an alliance. Overall, whether the option selected will or will not succeed will depend on the individual business model, the company and its staff, and its investors.

3.8.1.2 The Entrepreneur's Perspective

Entrepreneurs were asked what, in their opinion, a pharmaceutical company should offer to establish or enhance cooperations and networks with biotech start-ups. Overall, two general trends were noted. First, some entrepreneurs stated that the interest of major pharmaceutical companies in small biotech companies usually followed a cyclic behavior, determined by the size of their own cash flow or by the topicality of certain indications. According to the entrepreneur, a second distinctive of big pharmaceutical companies is their behavior, which sometimes appears opportunistically rather than strategically motivated. Despite these general trends, which the entrepreneur can hardly influence, it is obvious for most company founders that they have to establish ties to major pharmaceutical players fairly early on to:

- Learn how to best approach a company
- Obtain support in identifying those molecules that are most attractive for big pharma
- Profit from better drug development opportunities
- Profit from early selection of the most suitable drug candidates
- Get support in business development and securing IP rights
- Get support with clinical development expertise, complementary technologies, and marketing and sales
- Profit from cost/profit sharing
- Share networks
- Create international visibility.

Also, the perception of entrepreneurs is that the best way to meet representatives of pharmaceutical companies are partnering conferences, such as those hosted in partnership with the Biotechnology Industry Organization or the Novartis Venture Funds. One important need addressed by some entrepreneurs was for pharmaceutical companies they have entered into negotiations with to provide them with quick, honest, and transparent feedback and to clearly communicate decision-making processes and timelines. Seen through the eyes of the entrepreneur, this need is easily understood – considering that major decisions depend on this feedback.

3.8.2
The Role of Banks as Capital Investors

One question, posed to all three interview groups, was related to the role of banks as capital investors for biotech start-up companies. Overall, the role of banks as capital investors was considered marginally important from the perspective of entrepreneurs and even less so by venture capitalists and other investors. The role of banks as capital investors as perceived by the interviewees is displayed in Figure 3.8.

All interview partners agreed that the biggest hurdle banks faced when considering investing in high-risk biotech is their lack of expertise and technical know-how. Also, there was agreement that banks were not prepared to take on such high risks as those involved in biotech development, because they lacked the collaterals required to become active as investors. Nevertheless, banks do have some relevance during the development process.

Thus, for young companies having been granted early-stage government funds, banks frequently take on the role of cooperation partner of the government funding organization and act as brokers and trustees of the fund, the deficiency being guaranteed by the funding party. Also, in case of an exceptionally difficult situation turning up during development, it may become necessary for an entrepreneur to take up a short-term bank credit. In this case, it is certainly beneficial for the entrepreneur to already have established a good working relationship with his bank.

The rather marginal importance of banks as capital investors changes dramatically in the phase preceding an IPO, which banks may be very willing to support with mezzanine capital. Once companies have launched products on the market, debt financing may become important, particularly for the financial engineering of the balance sheet.

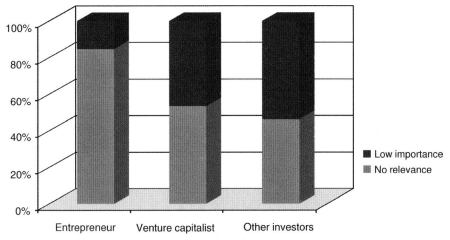

Figure 3.8 The role of banks as capital investors from the perspectives of the entrepreneur, the venture capitalist, and other investors.

3.8.3
The Importance of Milestone Payments

Payment obligations in license agreements can take the form of a one-time fee to upfront fees, technology access fees, annual payments, milestones, royalties, and option or extension payments. Milestone payments in a licensing agreement are payments made by the licensee to the licensor at specified times or when certain technological or business objectives have been achieved. Such incentives represent an important reward for an entrepreneur, or licensor, in return for sharing the opportunities of a specific development with the licensee. If milestone payments are intended to represent an advance to the licensor on future royalties, then such payments may be credited against royalty obligations. Alternatively, milestones are sometimes characterized as payments for prior or future research and development. To prevent conflicts over whether a milestone payment is due, milestone triggering events must be defined as clearly as possible (Ferruolo et al., 2002).

All three groups of interview partners were asked whether and how important it was to receive early milestone payments. There was no clear tendency within and between groups. Overall, more interview partners thought that milestone payments were moderately or highly important, even though they also acknowledged that it was rather difficult to achieve such early revenue streams.

The decision for or against milestone payments will have to be taken on a case-by-case basis and depending on the individual business model. One striking argument in favor of milestone payments is that they help motivate entrepreneurs to 'stay at it' and to maintain their commercial focus – an essential aspect for both corporate psychology and the entrepreneur, who has proven that he has taken his business model seriously. Conversely, milestone payments may also have a detrimental effect and damage the business model through hamstringing the company by setting and maintaining the wrong focus (Ferruolo et al., 2002).

3.9
Exit Routes

3.9.1
Exit Preferences

The questionnaire section on alliances in the late stage of a start-up asked all three interview groups what their preferred exit strategy for biotech companies was. The interviewees could choose one of the following options:

- IPO
- Trade sale
- MBO
- Other

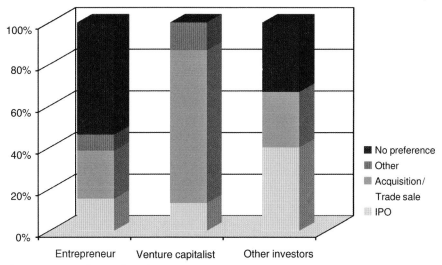

Figure 3.9 Exit preference from the perspectives of the entrepreneur, the venture capitalist, and other investors.

Whereas the majority of entrepreneurs did not seem to have a preference yet, venture capitalists clearly favored a trade sale exit. The group of other investors was fairly balanced, with three almost equally important subgroups (Figure 3.9).

Exit Preferences of the Entrepreneur

At the time the interviews were conducted, the goal of most entrepreneurs was not so much a clear exit vision but, rather, a general idea of the quality of the exit, namely that they would prefer an exit guaranteeing them the highest possible earnings and the highest level of independence inside the company. Their vision was to achieve a high value, which would then determine the possible exit scenarios. One of the most frequently mentioned scenarios other than those indicated in the questionnaire were partnering with a pharmaceutical company in form of a licensing agreement while at the same time staying independent.

Exit Preferences of the Venture Capitalist

For the venture capitalists, the most preferred exit route is the trade sale, mainly paid in cash, because it enables the investor to generate significant capital all at once. In contrast, an IPO is subject to a lock-up period of half a year before the shares can be sold. Also, the only share price determinant is market demand. However, an IPO also raises the public image and awareness of a company and moves it on the international stage, which might lead to a significant increase in market capitalization.

The optimal selling scenario will vary with the setup of each individual company. One of the key elements in the decision-making process is the number of drug candidates in the company's pipeline. Some venture capitalists also pointed out that an IPO often was not a true exit but, rather, represented another financing round for the company.

Exit Preferences of the Group of Other Investors

In the third group of interviewees, the decision which exit strategy to pursue again depended on the individual company profile and business plan. In this group, there was apparently no preference for a trade sale as it was the case in the group of venture capitalists.

3.9.2
The Role of New Markets

The next question was related to new markets, such as the Alternative Investment Market (AIM), and whether such markets provide an attractive alternative exit solution. AIM is the London Stock Exchange's international market for smaller growing companies. Since its launch in 1995, over 2900 companies have joined AIM – raising more than £34 billion in the process (www.londonstockexchange.com).

Overall, only few interview partners had already gathered some experience with AIM or similar new markets (Figure 3.10). Those who were familiar with AIM mostly stated they considered AIM as having too low a liquidity.

Therefore, there is still much insecurity when it comes to new markets such as AIM. Of the 2900 companies currently listed on AIM, only a minority is derived from the biotech area. Clearly, the decision might be different for a UK-based company that may find an opportunity to be listed on its home market very attractive. However, this 'home-country bonus' does not apply for companies outside the UK.

AIM does provide a market place for company presentations and gives early access to the capital market. Again, the final decision which stock market to choose will

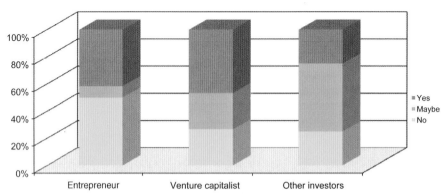

Figure 3.10 Are new markets, such as AIM, important for the entrepreneur, the venture capitalist, and other investors?

depend on the company's overall strategy, and the following questions should be addressed:

- How large is the market of my business?
- Where do possible investors look for investment opportunities?
- Where is the competition located and listed?
- How do I stay on the radar screen of the stock market?

3.9.3
Prominent European IPO Success Stories

One questionnaire item was dedicated to prominent success stories of biotech companies that had recently gone public, with multiple answers allowed. The majority of interview partners ($n = 28$) named Intercell (www.intercell.com), a vaccine company located in Austria, as a best-practice example, recognizing that this company had succeeded not only in setting up financing capital of almost unprecedented magnitude from investors all over the world, but also in showing an excellent after-market performance thanks to strong strategic alliances with other vaccines manufacturers, such as Novartis, GlaxoSmithKline, and Sanofi–Aventis. Some interview partners emphasized the role of the founder of the company, Professor Alexander von Gabain, who left his position as Head of the Department of Microbiology and Genetics at the renowned Research Institute of Molecular Pathology (IMP) at Campus Vienna Biocenter. Von Gabain founded Intercell and, after having acted as CEO for seven years, stepped back to open the way to Dr. Gerhard Zettlmeissl, an experienced CEO and well known in the pharmaceutical industry. Alexander von Gabain is still with Intercell, today acting as its CSO. Intercell's two most prominent pipeline candidates soon to reach the market are a Japanese encephalitis vaccine and a hepatitis C vaccine. More details on the success story of Intercell, including its financing and networking model, are given in Chapter 4.

Another example of a successful entrepreneurial venture mentioned by several interview partners is the Swiss-based company Cytos Biotechnology (www.cytos.com). Its founder, Dr. Wolfgang Renner, is still acting as CEO – an exception rather than the rule in the complex biotech start-up world. Cytos, too, is specialized in the development and manufacture of vaccines for chronic disease treatment and prevention. The application of its technology platforms allowed Cytos Biotechnology to build a full pipeline of what they termed Immunodrug™ candidates, targeting various chronic disease indications. Six vaccine candidates for treatment of nicotine addiction, allergy and asthma, Alzheimer's disease, hypertension, melanoma, and psoriasis are currently being tested in clinical trials (www.cytos.com).

Other Swiss-based IPO success stories mentioned were Actelion Pharmaceuticals Ltd (www.actelion.com), working in areas of high unmet medial need such as pulmonary arterial hypertension or Gaucher's disease, Basilea Pharmaceutica Ltd (www.basilea.com), focusing on new antibacterial and antifungal agents to fight drug resistance and on dermatology drugs, and Speedel Ltd (www.speedel.com), developing innovative therapies for cardiovascular and metabolic diseases. Other companies mentioned included the German Jerini AG (www.jerini.com) focusing on novel, peptide-based drugs and pursuing disease indications with as yet limited or no treatment options, German MorphoSys AG (www.morphosys.com) developing fully human antibodies for therapeutic, research, and diagnostic purposes, and the Italian start-up BioXell S.p.A. (www.bioxell.com) focusing on areas with unmet medical need in inflammation and urology. Other examples include the French company Cellectis (www.cellectis.com), a world leader in the research, development, and commercialization of rational genome engineering technology, and the Swedish company Biovitrum (www.biovitrum.com), which runs in-house research operations in the fields of blood diseases, pain, inflammation, obesity, diabetes, and other metabolic diseases requiring specialist care.

3.10
Primary Causes of Failure

Another question asked for the single most important reason why so many biotech companies failed. Some experts named two reasons, stating that these were the two most prominent and equally frequent ones (Figure 3.11).

3.10.1
The Entrepreneur's Perspective

From the entrepreneurs' perspective, the causes mentioned most frequently were the scientific risk involved in the development of any new biotech product and misman-

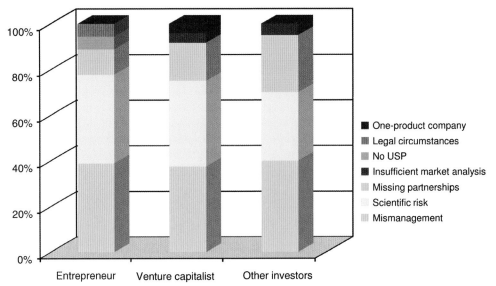

Figure 3.11 Primary causes of failure from the perspectives of the entrepreneur, the venture capitalist, and other investors.

agement. 'Scientific risk' was generally described as the risk involved in developing a new idea, technology failure, lack of proof of concept, problems with clinical trials, or having based the development project on the wrong scientific concepts. The term 'mismanagement' was described with phrases such as inadequately trained or inexperienced people, management failures, lack of economic orientation or management skills, inefficient management, and conflicts of interest.

Two entrepreneurs indicated that a lack of partnerships, e.g., with venture capitalists, could lead to failure of the biotech start-up, one interviewee mentioned lack of a USP from the business model, and one entrepreneur thought the legal circumstances, such as those regarding IP rights, were the main culprits for start-up failures.

3.10.2
The Venture Capitalist's Perspective

Venture capitalists likewise named scientific risk and mismanagement as the most important reasons for failure. The third most frequent reason was the lack of partnership with investors, leading to follow-up financing not being available, underinvestment, the inability to enter partnerships, and the inability to gain access to financial markets. One investor considered insufficient market analyses and another one-product strategies the most likely primary causes of collapse.

3.10.3
The Perspective of Other Investors

The group of other investors indicated mismanagement as the number one cause of failure, followed by scientific risk and lack of partnerships. One expert believed that an inadequately performed market analysis was the main reason for the ill-success of the company.

For entrepreneurs and venture capitalists, the results were similar with regard to the primary cause of failure, most likely because they are the groups mostly directly affected by such downturns. Most experts named mismanagement and scientific failure as the two most prominent and equally frequent reasons why many biotech start-up companies fail. For one interviewee, the minute difference between success and failure was timing, stating that in early stages management was the most important factor, whereas in later stages the scientific component was becoming more essential. In the group of other investors, mismanagement was mentioned ahead of scientific risk. This may be due to many experts in this group entering the funding process in a very early development phase, i.e., at a time when management failures are quite likely to occur.

Interestingly, unlike investors, entrepreneurs did not consider a lack of (financial) partnerships a factor potentially leading to failure of a young company. One reason for this may be that investors have, over the years, seen companies fail as a result of a wrong financing decision taken or a partnership with an investor entered into which failed to work out in the end. As one expert put it: "This will lead you into a vicious circle – if financing does not happen, management may be tempted to take the wrong decisions, making matters even worse." Overall, the reasons for failure given were homogenous between groups, with some particularities per group, which shows that both entrepreneurs and investors have a clear idea of why so many biotech companies remain unsuccessful.

3.11
Best-practice Countries for Funding Biotech Start-ups

All interview partners were asked which European country they considered the best-practice example for funding biotech start-ups. Some interview partners named two countries, indicating why in their opinion two countries with different entrepreneurial cultures were equally attractive. The results obtained in the three groups are displayed in Figure 3.12. The combined result for all three groups is shown in Figure 3.13.

3.11.1
The Entrepreneur's Perspective

Six of seven Austrian entrepreneurs stated that they considered Austria the best funding place for biotech start-ups. Whereas these answers reflected a clear bias, those of the other interview partners did not, with the countries they indicated being different from their countries of operation. The country mentioned almost as frequently by the remaining eight entrepreneurs was Switzerland ($n = 5$), followed by the UK ($n = 3$), Scandinavia ($n = 2$), Germany ($n = 1$), and Ireland ($n = 1$).

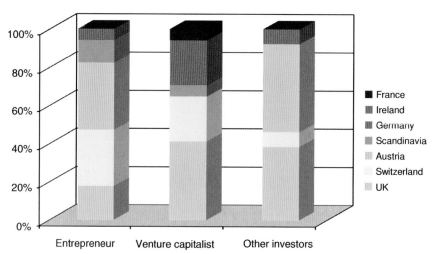

Figure 3.12 Best-practice country for funding biotech start-ups from the perspectives of the entrepreneur, the venture capitalist, and other investors.

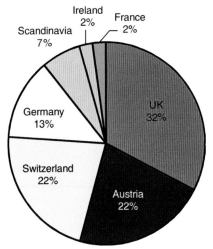

Figure 3.13 Best-practice country for funding biotech start-ups (all experts).

3.11.2
The Venture Capitalist's Perspective

The venture capitalists stated that the country they regarded as having the most advantageous start-up climate was the UK ($n = 7$), followed by Germany ($n = 4$) and Switzerland ($n = 4$). One vote each was given to Scandinavia and France.

3.11.3
The Perspective of Other Investors

For the group of other investors, the UK ($n = 5$) and Austria ($n = 4$) were the most start-up-friendly countries. Considering that four experts in this group were working for Austrian government funding institutions, the result that Austria was among the most attractive countries was not surprising. However, it was interesting to see the interviewees' unanimously strong belief in the UK biotech start-up climate.

The results across all three groups of interviewees indicate that the UK enjoys a high level of trust and is generally thought to offer a favorable climate for biotech start-ups, followed by Switzerland, Austria, and Germany. With the exception of Austria, this outcome appears to be fairly unbiased.

The arguments most frequently mentioned in favor of the UK market were:

- Cultural proximity to the US, the private-equity 'pioneer'
- Long-standing private equity and venture capital culture
- Early venture capital funding from university-based funds
- Angel networks
- Entrepreneurial culture
- Strong pharmaceutical clustering

• Target-oriented approach to research, with a high density of research biologists in the country.

The most prominent arguments in favor of Switzerland were:

• Access to large sources of capital
• Strong biotech, pharma, and research clustering
• High number of business angels
• 'Smart' universities
• Longer private equity and venture capital culture and the Anglo-American influence
• Entrepreneurial culture.

The most important argument in favor of Austria was the good public seed financing programs, even though the interviewees also pointed out that the financing challenges started to hit after the first round of funding. Some interview partners stated that Austria still had to develop its entrepreneurial spirit and its own private equity culture, while at the same time pointing to some prominent and successful scientists who had returned to Austria to pursue their research activities, such as Professor Josef Penninger, who worked with Amgen, the Ontario Cancer Institute, and the University of Toronto, Canada, for 13 years and relocated to Vienna in 2003 to head the Institute of Molecular Biotechnology of the Austrian Academy of Sciences (IMBA; www.imba.oeaw.ac.at).

The most important arguments in favor of Germany were the high level of scientific expertise and the availability of funding. This latter aspect was also seen in a rather critical light, however, with some interviewees stating that the past had seen too many investments in Germany and too few success stories.

The different groups of interviewees appeared to have different preferences with regard to the country with the most friendly climate for start-ups. Switzerland's attractiveness was primarily recognized by the groups of entrepreneurs and venture capitalists, whereas Germany was favored by the groups of venture capitalists and other investors. Overall, this suggests that Austrian and German entrepreneurs consider the entrepreneurial start-up conditions in Switzerland more favorable than in Germany.

3.12
Most Striking Differences between the US and European Venture Capital Markets

The last questionnaire item asked what the interviewees thought were the most striking differences between the US and European venture capital markets and how they perceived these differences.

The differences between the US and the European venture capital markets as seen by the entrepreneur, the venture capitalist, and other investors are presented in Figure 3.14. The possible differences as perceived by all three groups of interviewees taken together are presented in Figure 3.15.

The most prominent difference given by interview partners across all groups was the size of venture capital funds available for process and exit funding. In terms of the

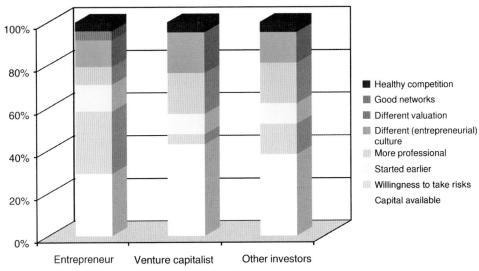

Figure 3.14 Major differences between the US and the European venture capital markets from the perspectives of the entrepreneur, the venture capitalist, and other investors.

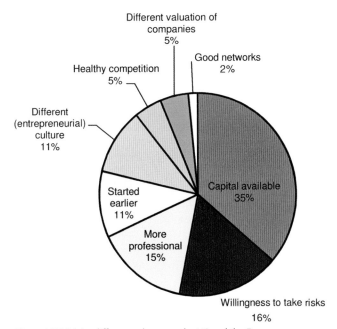

Figure 3.15 Major differences between the US and the European venture capital markets (all experts).

second most important difference, the perceptions of entrepreneurs and investors showed rather striking divergences.

For the entrepreneur, one key difference between the US and European venture capital markets was the willingness to take risks, which they believed was much less pronounced in Europe than in the US. In contrast, only one venture capitalist and three other investors thought that the two venture capital markets differed in terms of their risk-taking capacity. This suggests that entrepreneurs believe that European investors are not prepared to take the same level of risk as their US counterparts. This observation may be related to another difference pointed out by some interviewees from all three groups, i.e., the healthy competition that prevails in the US. Competition among US venture capitalists is fiercer than in Europe, potentially leading to quicker decisions carrying higher risks.

One interview partner put it this way: "In the Anglo-American culture, taking risks means taking chances. The German way of thinking is to translate risk into danger." One entrepreneur stated that many European venture capitalists were too preoccupied with amassing money. A further aspect, which may be related to the Americans' willingness to take risks and that was mentioned by one entrepreneur and two venture capitalists, is that one and the same company, valuated both in the US and in Europe, will achieve a significantly higher valuation in the US.

The difference ranking third in the overall analysis is that US venture capitalists have a more professional approach to work thanks to their higher level of experience. As a result, they are also more confident in evaluating companies and their projects. This opinion was particularly pronounced among venture capitalists and other investors, but played less of a role among entrepreneurs.

Not surprisingly, the fourth major difference is that the US venture capital market can look back on a longer history than its European counterpart. Also, the entrepreneurial cultures seem to differ remarkably. Thus, the US is believed to be the champion in identifying and implementing innovative ideas, and entrepreneurs in the US usually try to make their companies come out big. The 'American dream' that anyone can rise from rags to riches is alive today, as is the belief that the US still is the land of boundless opportunities. In contrast, as some interview partners pointed out, Europe does not have a culture of failure. Even worse, failure is stigmatized and must therefore be avoided at all costs. In the American culture, failure is a part of (economic) life and believed to lead to key learnings that can later be harnessed in pursuing new business opportunities and in trying to succeed again.

Also, some interviews showed that there appears to be a difference in how negotiations are conducted. In the US, discussions and negotiations are usually tougher and more target-oriented, taking financial figures for what they are worth, a characteristic reflecting the dominant role US venture companies usually play on the Boards of Directors of the companies they finance. In Europe, the Boards of Directors are more entrepreneurially friendly and more lenient when it comes to taking decisions.

Related to the longer tradition and higher level of experience of the US venture capital market is their integrated value chain, covering the entire development process of a newly founded company. First, the US has a high number of professionally organized tech transfer offices installed at universities, helping young scientists with a great and innovative business idea found their own start-ups. The primary financing source for many such companies in this early phase are business angels, individuals who may have been entrepreneurs themselves and now find it fascinating to make their know-how and experience available to younger colleagues. They are frequently organized in private networks. They have the knowledge, the experience, the network, and the money – all factors that any new entrepreneur is highly dependent upon. After that, the venture capitalist will step in, likely to act quicker than the European investor in order not to lose any major opportunity to a competitor.

Overall, therefore, there are still some major differences between the European venture capital market and its 'older brother' in the US, with Europe still in a catch-up phase. However, there was also agreement among interviewees that the past decade has seen the gap between the European and US venture capital markets become substantially smaller.

4
Biotech Case Studies

This chapter presents in-depth interviews with five entrepreneurs from the Austrian biotech scene. These case studies are meant to support and corroborate the results obtained in the structured qualitative interviews with entrepreneurs, venture capitalists, and other investors presented in previous chapters.

All five entrepreneurs were asked the same questions in the same order to enable a comparison between the different start-ups. All companies are located in Austria, where a small but vivid biotechnology landscape has developed in the past ten years. The following examples demonstrate how even in a small country with a culture that is less private-equity and entrepreneurially driven, entrepreneurs succeed in building alliances and global networks to finance their start-ups.

The entrepreneurs were selected according to the following criteria, representing:

- a good mix of companies in different development stages, one in seed, two in early, and two in late stage development
- a mix of entrepreneurs having applied different financing approaches
- a mix of entrepreneurs who developed their company as a spin-off from an academic institution or another pharmaceutical company
- a high level of expertise in the field of biotechnology/pharmaceutical drug development and/or business management.

4.1
AUSTRIANOVA Biomanufacturing AG

Interview with Mag. Thomas Fischer MBA, CFO, and Dr. Susanne Bach, CC
— 31 March 2008

Thomas Fischer, one of the founders of AUSTRIANOVA Biomanufacturing AG (www. austrianova.com), has been with the company since its inception in 2001. He has more than 17 years of experience in Business Management, Finance, and Controlling, with 10 years spent in biotech and pharmacy. Thomas Fischer started his career as Group Controller for the Gusta Food Group and then joined the Nycomed Pharma Group as Director of Finance and Controlling. Before founding AUSTRIANOVA, Thomas Fischer had been Commercial Project Manager for Aventis Crop Science.

Biotech Funding Trends: Insights from Entrepreneurs and Investors.
Alexandra Carina Gruber
Copyright © 2009 WILEY-VCH Verlag GmbH & Co. KGaA, Weinheim
ISBN: 978-3-527-32435-4

Susanne Bach is Head of Corporate Communications at AUSTRIANOVA. She started her career in biomedical academic research. After having completed postgraduate training in public relations and communications, she turned to the pharmaceutical industry. She served as communications and new media relations manager at Pfizer Austria for four years before joining AUSTRIANOVA.

Question 1 – When and how was the biotech start-up founded?

▶ "AUSTRIANOVA Biomanufacturing AG was founded in 2001 as the first spin-off from the University of Veterinary Medicine in Vienna. Our vision was to develop proprietary delivery technologies for health care applications and to validate them with a marketable reference product. We currently have 32 full-time employees."

Question 2 – How has your company developed since its foundation and what products are you currently working on?

▶ "Since its foundation in 2001, AUSTRIANOVA has steadily developed from an academic start-up to a manufacturing-focused, internationally recognized technology leader and one of Austria's most important biotech start-up companies next to Intercell. AUSTRIANOVA is currently the world's only company to focus its research on what is called 'bioencapsulation,' the encapsulation of living cells for therapeutic purposes in the development of new anti-cancer therapies (Figure 4.1). Representing the most important strategic development area of AUSTRIANOVA, our reference product NovaCaps® is the first encapsulated cell product targeting solid tumors that are difficult to treat."

"This project stimulates and fuels the continued development of our company. NovaCaps® has received orphan drug designation from the European Medicines Agency EMEA for what is currently the most poorly treatable oncological indication – pancreatic cancer. The name of the product is NovaCaps® Pancreas. Having been granted orphan drug status has significantly speeded up the regulatory pathway toward marketing authorization and guarantees a ten-year market exclusivity period after approval by the EMEA. In this indication, the product is currently in pivotal Phase III trials."

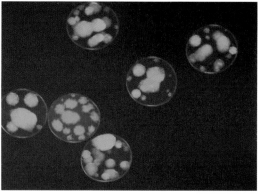

Figure 4.1 Encapsulated living cells under the fluorescence microscope.

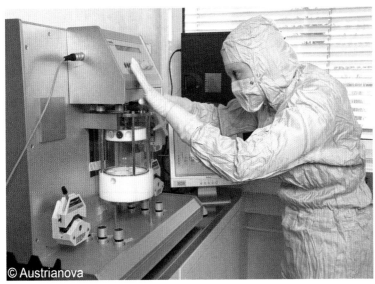

Figure 4.2 NovaCaps® Encapsulator – GMP-compliant manufacturing of encapsulated cell products.

"Moreover, AUSTRIANOVA has developed a unique IP-protected Good Manufacturing Practice (GMP) production line for the encapsulation of living cells in cellulose sulphate that can be applied to a wide range of therapeutic applications (Figure 4.2). The GMP production facility in Frankfurt, Germany, enables the consistent and quality-controlled encapsulation of living cells for clinical applications at an industrial scale, not only for our reference product NovaCaps®, but also for other living cells encapsulated by the same technology. Bioencapsulation therefore represents our core asset. We are currently working on further expanding the NovaCaps® approach to other poorly treatable locally occurring solid tumors, such as liver cancer or head and neck cancer."

Question 3 – How has your company tackled the financing challenge?

▶ "We are proud to say that AUSTRIANOVA relies on a solid financial background. This enables us to enter active partnerships with major players from the pharmaceutical and biotech areas, opening up further interesting co-development options to put the bioencapsulation technology to further use."

"In early 2007, we were delighted to announce that the Irish Ryan Group Holdings under their company founder and CEO, pharmacist Gerard Ryan, made a substantial participation in our company. These funds, amounting to an overall volume of €35 million in a Series B financing round, will enable us to realize the full potential of our proprietary NovaCaps® in pancreatic and other cancer indications and will be implemented in several steps. In two previous financing rounds between 2002 and 2004, we were able to raise a total of €30 million from several Austrian and international investors by putting together an innovative mix of capital, grants, and various loans. Through the intelligent use of a number of diverse financing forms we were able to bypass the classic financing option venture capital. Together with our new partner, Ryan Holdings, we entered into a new development phase, financing additional partnering projects to develop other therapeutic approaches based on the cell encapsulation technology."

Question 4 – How important are cooperations?

▶ "From my perspective, cooperations do play a very important role. They enable us to get strategic access to partners potentially enabling us to further develop and commercialize our cell encapsulation technology for a wide range of new therapeutic applications, even outside the field of oncology. Our partners for the industrial production of NovaCaps® are Miltenyi Biotec GmbH and the Fraunhofer Institute for Applied Polymer Research IAP."

"Last but not least, from a financial perspective, the fund raising through Ryan Group Holdings will enable us to realize the full potential of our proprietary NovaCaps® technology in pancreatic and other cancer indications and to further exploit new co-development options with major pharmaceutical and biotechnology players."

Question 5 – What are the key challenges and lessons learned when it comes to financing and looking for alliances?

▶ "As one of our key learnings, we had to acknowledge that the areas of pharmacy and biotechnology are part of a highly structured, commercially oriented, and regulated market with high entry barriers and clearly defined rules."

"From my perspective, one of the major challenges for an academically oriented spin-off is to start, as early on as possible, to detach and dissociate oneself from the scientific working environment and to adapt to the requirements of the market. The same holds true for questions related to financing. You quickly have to get used to explaining your technology to investors and to translate for them what future benefits they can expect."

Question 6 – What will be the next steps for AUSTRIANOVA Biomanufacturing AG?

▶ "As a first step, we are eager to close new co-development agreements with other pharmaceutical and biotech companies to further stimulate the organic growth of AUSTRIANOVA into a biopharmaceutical manufacturer. Another important goal for us is to develop into an economically oriented company while at the same time maintaining the dynamics of a small or medium-sized enterprise."

4.2
AVIR Green Hills Biotechnology AG

Interview with Mag. (FH) Michael Tscheppe, CFO, Mag. Reinhard Zickler, COO, and Isolde M. Bergmann, CC and HR — 2 April 2008

Michael Tscheppe founded AVIR Green Hills Biotechnology AG (www.greenhillsbiotech. com) in 2002 together with Thomas Muster, at that time Head of the Viral Oncology Laboratory at the Department of Dermatology, Vienna Medical University and contributor to several inventions providing the basis for the company's platform technologies. Michael

Tscheppe started his business career working for a marketing agency, then held various positions at the Austrian Post AG, where he gained his profound experience in finance, marketing, administration, and organization. Michael Tscheppe holds a degree in European Business Management.

Reinhard Zickler first met the inspiring company founders in late 2003 and joined the company as COO in 2005. He started his business career as manager in regional and international real estate project development, where he acquired broad financial, organizational, project management, and business development skills. He is part of a tight network of high-level national and international contacts with executives and decision makers in diverse fields of business. Reinhard Zickler holds a master's degree in law.

Isolde M. Bergmann is in charge of Corporate Communications and Human Resources. After studying pharmacy at the University in Vienna, she later switched to the areas of marketing and sales. Isolde Bergmann obtained wide experience in Marketing and Human Resources in one of Austria's major travel agencies for a total of 12 years, acting in an executive position for five years before joining AVIR Green Hills Biotechnology in February 2006.

Question 1 – When and how was the biotech start-up founded?

▶ "AVIR Green Hills Biotechnology AG was founded in 2002. The company is based in Vienna, Austria, and currently employs more than 50 associates. Our products represent unique therapies against viral infectious diseases and cancer designed to significantly improve human health and quality of life."

Question 2 – How has your company developed since its foundation and what products are you currently working on?

▶ "The core competence of AVIR Green Hills Biotechnology is our extensive know-how in the field of virology, with special emphasis on the interactions between viruses and their host cells. This expertise has allowed us to establish a 'viral engineering' product line, which has resulted in the development of two products currently in late-stage preclinical development and a portfolio of earlier-stage programs in the pipeline."

"Today our lead project is FLUVACC. This live attenuated replication-defective influenza virus vaccine, brand-named deltaFLU, is the first product to emanate from our viral engineering technology (Figure 4.3). DeltaFLU represents a live attenuated influenza virus with an optimal balance between attenuation and immunogenicity. It is generated by reverse genetics and is produced in mammalian cells. This allows fast and efficient generation and production of vaccine strains against influenza. DeltaFLU is applied intranasally with a spray device."

"Compared to the traditional inactivated virus preparations currently dominating the influenza vaccine market, live attenuated viruses induce longer-lasting and more cross-protective immunity with significantly higher protection rates. We believe that, due to significantly higher efficacy rates and the ease and painlessness of intranasal administration, the live attenuated approach is superior to the inactivated injectable flu vaccines currently on the market. Although the live attenuated viruses on the market have a good safety record, shedding of the vaccine virus does occur, and unexpected complications in young children, the elderly, and immuno compromised persons might arise from widespread use."

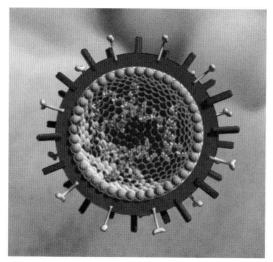

Figure 4.3 Influenza virus.

"Constructing a vaccine virus which undergoes only abortive replication and at the same time is capable of inducing a strong immune response eliminates the limitations and complications associated with traditional live attenuated virus vaccines – and this is the core concept of our FLUVACC delta NS1 technology."

"On this basis, we have generated live attenuated replication-defective viral vaccines and found that influenza A and B viruses based on the FLUVACC technology are highly attenuated in animals, such as monkeys and ferrets, and can provide protection from challenge with wild-type viruses. The particular properties – apathogenicity due to abortive replication while being highly immunogenic – make deltaFLU attractive as a safe and effective vaccine against endemic and pandemic influenza."

"Containing influenza epidemics, and above all bird flu, is a priority in many countries' health policies, and both the US and the EU bet on quickly developing an efficient human vaccine. The EU opted for our innovative vaccine development delta NS1 technology as a future European pandemic preparedness strategy. Even though the vaccine looks like a pathogenic influenza virus to the body, it does not cause disease because its pathogenicity factor NS1 was deleted. As a result, after intranasal application, the vaccine stimulates a strong immune response affording protection against influenza without causing illness of the host."

"In 2007, we also received 'indirect confirmation' of the importance of our FLUVACC development program, when Thomson Pharma®, a comprehensive global pharmaceutical information solution that covers the entire drug discovery and development pipeline, ranked our product among the five most promising drugs entering Phase I trials (Table 4.1)."

"Our second product in development is MelVir. It is based on our findings whereby melanoma is associated with the expression of endogenous retroviruses, an observation enabling early diagnosis and therapy of melanoma."

Table 4.1 The five most promising drugs entering Phase I trials (Thomson Pharma®, 2007).

Drug	Disease	Company
Ad 35 HIV-EnvA vaccine	HIV infection	GenVec/NIAID Vaccine
Nestorone and estradiol (transdermal gel)	Female contraception	Antares/Population Council
FLUVACC	Influenza	AVIR Green Hills Biotechnology
Insulin oral capsule	Diabetes	Oramed
Trodusquemine	Obesity	Genaera

"Retrovirus-specific antibodies are prevalent in more than 90 percent of melanoma patients. Importantly, antibodies can be detected at an early stage of the disease, which is vital for the cure of melanoma – the central finding of the development of our MelVir diagnostic tool."

"During our research, we identified a new melanoma-associated antigen, i.e., melanoma-associated endogenous retrovirus, or MERV. Specifically, we showed that MERV-derived antigens are expressed in primary melanomas and metastases but not in benign lesions, suggesting that expression remains stable throughout the entire process of cancer progression. Moreover, the finding that melanoma patients have circulating antibodies that are reactive with antigenic determinants of the MERV envelope protein indicates that these retroviral proteins are immunogenic. MelVir diagnostics and MelVir therapeutics are based on this key finding."

Question 3 – How has your company tackled the financing challenge?

▶ "The first years after our company foundation in 2002 were financed with committed government funds amounting to approximately €2.5 million. Particularly the year 2002, marked by the breakdown of financial markets worldwide, was a hard time for us and many other biotech start-ups. Starting a biotech company at that time was not an easy thing to do. Despite these unpleasant external conditions, we were extremely successful in raising public money from Austrian government institutions and the EU."

"In 2005, another important milestone was achieved when we, together with highly renowned research partners, were granted EU funding of €9.3 million from the Sixth Framework Program (FP6), the financial instrument created to make the European Research Area (ERA) a reality. Projects benefiting from this program have to be transnational, meaning that only consortia of partners from different member states and associated countries can apply. Proposals are evaluated and selected for funding by the European Commission with the help of independent external experts. Evaluation and selection of proposals is a highly competitive process: fewer than 20% of all proposals are funded. Hence, selected projects represent the top proposals of European research, and we were very proud of having been appointed lead participant and coordinator for four projects, two of them particularly designed for pandemic influenza development."

"Currently, we are working on a new financing round with a private placement that should enable a capital increase of €15 million in order to reach our next important milestone, a challenge study supposed to deliver results in 2009. The primary purpose of the study, conducted in the UK, is to determine whether our influenza vaccine prevents the development of influenza symptoms in human volunteers. Each volunteer is given either the vaccine or placebo and will then be exposed to, or 'challenged with,' a live influenza virus. A Phase II study is scheduled to start in late 2009 or early 2010. Generally speaking, results of studies of antiviral or vaccine approaches in the challenge model have been predictive of efficacy. A successful outcome in the challenge model will certainly help us to significantly reduce the risk of the product in subsequent development stages."

"Overall, we are convinced that we are quite unique in our financing approach because more than five years after the foundation of the company, its founders still hold the majority of shares (60%). So far, our investment strategy has relied on family and friends, atypical silent partnerships, and national grants from Austrian public funding institutions and the EU – without participation of a venture capitalist. Our strategy is to finish the indicatory challenge study in 2009. Afterwards, depending on the results and the feedback from our existing investors, we will evaluate which steps to take next. Options range from a new financial round with existing or new investors to searching for a strategic partnership with a pharmaceutical company. We may also consider taking a venture capitalist on board later on, although so far our goal has been to protect the company founders' interest as long as possible to give them the unique opportunity to be the ones to profit most from the company's success without being diluted by venture capital participation. Also, we have successfully attempted to obtain an ambitious valuation, which was recently confirmed by one of the biggest players in the pharmaceutical M&A business. If the results of the challenge study are positive, the development risk of this vaccine will from then on be significantly lower, representing a major milestone in this still early development phase."

Question 4 – How important are cooperations?

▶ "Cooperations have played an important role from the very outset and will definitely continue to be important in the future. Thanks to the open-minded and inquisitive spirit of its founders, AVIR Green Hills Biotechnology has been able to profit from many scientific cooperations and networks set up across Europe. Moreover, we have always been prepared to outsource specific tasks, such as GMP production, as well as to take additional competencies on board in case they were outside of our own core competencies. We believe that, over time, it pays off to always search for specialists in a particular area and to try to establish some sort of partnership with them."

"A major challenge was certainly to find a contract manufacturer for the production of the medication needed for our clinical trials. The potential for automation in the conventional production in embryonated chicken eggs is limited. Also, this traditional method requires the availability of pathogen-free, embryonated chicken eggs, which poses a problem, particularly in times of a pandemic. Therefore, we developed a production system in Vero cells, a continuous cell lineage that can be replicated through many cycles of division. This production system permits production in bioreactors, automation, and manufacture at any scale. The technology comprises not only the novel, more efficient vaccine, but also an innovative production method in Vero cell cultures as well as a fast and efficient method, i.e., reverse genetics, for producing the vaccine viruses."

"With regard to other cooperations, we are proud of leading an international consortium comprising several renowned national and international research partners, among them small to medium-sized enterprises and academic institutions, such as Vienna Medical University, the German Robert Koch Institute, Goethe University Frankfurt, and Retroscreen London. Also, as indicated earlier, we are a lead participant and coordinator of four EU projects run under the Sixth Framework Program in a cooperation with highly renowned research partners."

"For Phase III clinical trials, regulatory approval, and marketing and sales, we will seek strategic partnerships and cooperations with pharmaceutical or biotechnological companies. We have already been able to establish close cooperations, but we will also continue our networking efforts in the next years. Moreover, we are keeping contacts to some venture capitalists who we might consider taking on board if necessary. Obviously, this is not our main financial goal, but an option we consider. As mentioned, we want to protect the interests of the company founders as long as possible without being diluted. Going public is also an important topic and is planned for 2009 and 2010 if the company's development is doing well and our results will allow a good performance."

Question 5 – What are the key challenges and lessons learned when it comes to financing and looking for alliances?

▶ "One of the major challenges was definitely to go against the flow of biotech funding, which, for most biotech start-ups, is still strongly tied to venture capital. In addition, in a small country like Austria, venture capital sources are scarce, so that, from the very beginning, you have to keep looking for funds across Europe or even beyond. Moreover, with financial markets having hit rock bottom in 2002, it was almost impossible to get a fund to invest in a young biotech start-up."

"We therefore made a virtue of necessity and chose to take an untrodden path. So far, things have worked out very well. We have been able to rely on several investors from our immediate environment of family and friends, most of them not related to the biotech area. We are quite pleased that these investors and us have, over the years, grown into a 'big family.' This may in part be due to our having realized from the beginning how essential it is to identify those investors who are fully in line with our philosophy and who are convinced of our innovative technologies. Today, having successfully passed Phase I clinical trials, we know that our approach is a feasible model, that it really works, and that it will pay off in future."

"With regard to some of our key learnings, we are convinced that having complementary skills on board is essential, no matter which development stage you are in. In our case, for example, the supervisory team is made up of top-class, highly experienced experts with a long track record in various industries, who have contributed a lot of constructive ideas and provided diverse viewpoints. Also, we maintain very good relationships with our investors. We all share one vision, and everybody is very enthusiastic, excited, and highly motivated to be a part of the success story that is taking shape. In our opinion, the primary asset of a biotech start-up is certainly the human capital. You have to recruit the best person for each job, to incentivize them, and to keep them motivated so they will stay in the company as long as possible. At AVIR Green Hills, we place special emphasis on the 'human factor' and on the 'soft skills' of our team, because these can contribute greatly to the overall success of the company. We have a team of highly specialized, qualified, and motivated people who have been taking a significant amount of risk to follow us and our vision."

"Another key learning is that you should never set up your development program without taking into account possible delays or setbacks. Therefore, always be sure to develop a backup strategy, because you cannot expect everything to work out the way you had envisioned. We have been fortunate enough to have had most things turn out as planned, but you should never take such scenarios for granted."

"Finally, the development program of a vaccine differs substantially from that of other biological products, such as monoclonal antibodies. Vaccines have some specific characteristics as well as some major advantages, particularly regarding a significantly lower risk after successful completion of Phase II clinical trials. Therefore, we are confident and excited about the upcoming start of our Phase II study and the subsequent challenge study."

Question 6 – What will be the next steps for AVIR Green Hills Biotechnology AG?

▶ "The first and most important step will be the outcome of our upcoming clinical studies, which will automatically determine the path we will take in the next financing round. Depending on the results, we will see whether our current investors will invest again or whether we will have to look for a new investor. After successful completion of the Phase II study with deltaFLU, we might be looking for a strategic alliance with a pharmaceutical company or even for an IPO. This decision will depend not only on our own company development, but also on external factors, such as the future development of the financial markets."

4.3
EUCODIS GmbH

Interview with Dr. Wolfgang Schönfeld, CEO — 31 March 2008

Wolfgang Schönfeld is the founder of EUCODIS GmbH (www.eucodis.com). Before that, he served as President of MIXIS France after its acquisition by PLIVA d.d. in 1999. At that time, he was also Director of Anti-infective Research at PLIVA, the largest Central European pharmaceutical company, headquartered in Zagreb. He started his industrial career at Bayer AG in 1991, holding several positions in pharmaceutical research and management. Wolfgang Schönfeld is a physician by training and obtained his MD at the RWTH Aachen, Germany. He later became lecturer for experimental microbiology at Ruhr University, Bochum, where he spent seven years doing post-doctoral research in immunology and microbiology.

Question 1 – When and how was the biotech start-up founded?

▶ "EUCODIS GmbH was founded in May 2004 in Vienna, Austria, as a spin-off from MIXIS France, a biotechnology company belonging to the PLIVA group, the largest Central-Eastern European pharmaceutical company in Zagreb, Croatia. The rationale and underlying scientific concept for the creation of our company is to adapt the mechanisms of evolution to the generation of novel products for the life sciences. Our unique proprietary technologies, *in vivo* recombination and somatic hypermutation, allowed us to organize our business around two core areas – the generation of novel pharmaceuticals, generally referred to as 'red biotech,' and the development of industrial enzymes and biocatalytic processes, known as 'white biotech.'"

"We exclusively licensed all IP and know-how and took over all employees from MIXIS as well as the premises at Necker Faculté de Médecine at Paris. In the months that followed, business operations were put in place, and the company headquarters and a research site were established on the Novartis research campus in Vienna, Austria."

"In addition to the headquarters and main laboratories in Vienna, we established a subsidiary at one of the leading European research institutes, Necker Faculté de Médecine in Paris. Our staff size was about 28 associates, led by an experienced management team which had previously worked together for several years and was comprised of individuals with profound knowledge and skills in both pharmaceutical research and management."

Question 2 – How has your company developed since its foundation and what products are you currently working on?

▶ "From the beginning, we have been able to translate our scientific success into business: Several industrial collaborations were launched, according to our business strategy mainly in the field of white biotechnology, proving the power and superiority of the *in vivo* recombination technology. Collaboration partners such as Sanofi–Aventis, Sandoz, Henkel, Lohmann Animal Health, and PLIVA showed that the market accepted this technology. EUCODIS is currently working on optimizing several enzyme classes, and significantly improved enzyme candidates for further development by our partners have already been generated. In parallel, we were successfully launching three in-house research projects for the generation of novel biopharmaceuticals."

Development of a C5a antagonist

"The goal of our C5a antagonist project is to generate a novel anti-inflammatory peptide based on the inhibition of the complement factor C5a for treatment of severe systemic inflammatory conditions. Using somatic hypermutation, we aim at generating a variant with C5a-antagonizing properties. The project was submitted to the Vienna Life Science Call in early 2004, reviewed by an international expert team, and awarded first prize in a robust competition of 30 applicants."

Generation of a novel bone factor for the treatment of fibrodystrophia ossificans progressiva

"Fibrodystrophia ossificans progressiva (FOP) is a rare hereditary disorder, characterized by the production of ectopic bone in muscle tissue, in turn causing progressively increasing disabilities in mobility and ultimately leading to death of the patient as a result of respiratory insufficiency. No therapy is currently available. Using our technologies, we seek to develop an antagonist to become the first therapy option for FOP patients. The project was submitted to the Vienna Fempower Call in 2004 and won third prize in a peer-reviewed competition."

Affinity maturation of a human monoclonal anti-inflammatory antibody

"Various inflammatory disorders, such as rheumatoid arthritis, asthma, psoriasis, and transplantation rejection, are characterized by an overproduction of certain cytokines. Starting with an existing human antibody having weak affinity for an inflammatory cytokine, the somatic hypermutation technology is applied to generate an improved human antibody with very low or no antigenicity and a therapeutically suitable neutralizing capacity. The technology has matured to industrial applicability, and the first antibodies are currently being optimized by the technology. The research project was heavily supported by a grant from the Research Promotion Agency FFG of the Austrian government, the central federal institution for the enhancement of research, development, and innovation in Austria."

"To bolster our pipeline in pharmaceuticals, we decided to license in an advanced clinical project in the indication of breast cancer. The therapeutic approach is based on the inhibition of the key enzyme for estrogen production, aromatase, using the safe and potent inhibitor formestane. In previous studies, topical application demonstrated good tolerance, significant absorption, and in a compassionate use trial high efficacy in previously untreated newly diagnosed breast cancer patients. We originally licensed the project from our US partner MDI in 2006 for EU markets and in 2007 decided to fully acquire the project and worldwide rights. We obtained ethics committee approval for conducting a Phase IIa study, which is scheduled to start in the second quarter of 2008."

"The success in both areas – pharmaceuticals and industrial enzymes – led to EUCODIS gaining significant visibility in the biotech scene. In early 2006, we therefore decided to develop the company in the form of two independent sister companies, EUCODIS Bioscience, an entity focusing on white biotechnology, and EUCODIS Pharmaceuticals. In September 2007, the separation of EUCODIS was finalized, together with a successful Series B financing round for EUCODIS Bioscience."

Question 3 – How has your company tackled the financing challenge?

▶ "Since its foundation in 2004, EUCODIS has relied on four principal sources of income, venture capital, industrial collaborations, grants, and loans, so far having raised a total of about €8 million."

Venture capital

"In November 2004, a Series A round of €1.8 million was closed, led by Pontis Venture Partners, Vienna, with Gamma Capital Partners as co-investor. Further investments up to a total of €3.5 million were done during 2006 and 2007. In September 2007, i.e., during the separation of EUCODIS into two separate entities, we successfully closed a Series B round for EUCODIS Bioscience amounting to €3.2 million. The syndicate was led by Athena Beteiligungen AG, Vienna, with the participation of the German Acceres GmbH, Pontis Venture Partners, Gamma Capital Partners, and private investors."

Industrial collaborations

"In terms of industrial collaborations, our goal was to rapidly demonstrate the power of our recombination technology and the acceptance we received from the market. For strategic reasons, we started marketing our technology in the field of 'white biotech,' generating significant income of about €1 million within the first three years. Our clients and partners are Sanofi–Aventis, Sandoz, Lohmann Animal Health, GlaxoSmithKline, and Henkel. These collaborations were transferred to EUCODIS Bioscience, because their activities were fully centered on industrial enzymes."

Grants

"EUCODIS has received various non-refundable grants by peer-reviewed calls of the ZIT Center for Innovation and Technology, among them the previously mentioned:

- First prize in the Life Science Call in 2004 for our C5a antagonist project, a potent anti-inflammatory target
- Third prize in the ZIT Fempower Call in 2005 for identification of a novel therapeutic principle for FOP
- Third prize in the ZIT Cooperate Call for a project carried out in cooperation with PLIVA d.d. for the identification of novel enzymes for macrolide synthesis."

"In addition, we have successfully applied for support by the Austrian FFG, receiving a substantial grant-loan combination for the further development of the antibody technology platform. We participate in a major collaborative research grant, the Human Monoclonal Antibodies for a Library of Hybridomas (HYBLIB), that is part of the Seventh Framework Program for Research and Technical Development (FP7); together with four academic institutes, we are working on extending our antibody technology platform. The program scored third among all applicants and won a total of €2 million. Finally, together with our partner at the University of California, Los Angeles (UCLA), we successfully applied for a National Institutes of Health (NIH) grant for a novel class of antibiotics based on a rationale developed at UCLA and on the EUCODIS *in vivo* recombination technology."

Loans

"The start-up in May 2004 was financed through a seed-financing loan granted by the AWS, a credit line allowing us to set up the company and its operational structure. In the course of the licensing-in of the breast cancer project, EUCODIS successfully applied for a double-equity credit line in 2006 that was guaranteed by the AWS. The grant-loan combination mentioned earlier consists of 50% funding and a 50% long-term loan. A major loan was granted within what is called the Technology Financing Program (TFP) of the AWS, a program established to support the development of young, small- and medium-sized companies. This loan enabled us to start research on the FOP project."

Question 4 – How important are cooperations?

▶ "Since its foundation in 2004, EUCODIS has set up various collaborations and partnerships in three areas:

- Industrial partnerships, where we carry out projects geared towards providing our partners with the products they need
- Academic collaborations to broaden and improve our existing technology platform

- Active participation networks between industry and academia to remain at the cutting edge of market developments and technologies."

Industrial partnerships

"The years between 2002 and 2004 were a difficult period for financing biotech newcomers. In contrast to previous years, we were faced with a profound shake-up of the world's finances, and following the burst of the dot.com bubble, investors were extremely cautious about investing in new technologies, preferring projects that were already up and running, especially advanced preclinical projects. Therefore, from the very outset our strategy was to profile ourselves among industrial leaders as a recognized partner, to develop our own project pipeline as fast as possible, and to eventually license in an advanced project. Overall, this strategy has worked out successfully, as indicated above. The industrial partnerships and collaborations have played and will continue to play a key role in both EUCODIS Bioscience and EUCODIS Pharmaceuticals:

- Collaborations offer significant income for our company in addition to funds raised through grants, loans, and venture capital money. This is especially true in 'white biotech,' a market dominated by projects that generate short-term income, are technology-driven, and have a significantly better risk profile compared to pharmaceutical projects. However, the total earnings are also significantly lower than with pharmaceuticals.
- Collaborations train the entire company to develop a customer-oriented approach. Young researchers, most of whom started their industrial career in EUCODIS, learn to work in teams covering a broad span of technologies and know-how. In most cases, team members will have to closely interact with their colleagues from the partner company, exposing them to different cultures, approaches, and expectations that have to be dealt with cultural empathy to ensure success of the project.
- Collaborations introduce a sense of urgency into the team, they secure critical feedback on one's own capabilities, and have thus created a project-driven approach within EUCODIS.
- The interaction with colleagues from other companies is fundamental for our management, enabling us to identify the best strategy for our own company and projects. The close interaction with market participants influences the development strategy of each of our projects and, in turn, increases the value of our technologies."

"The above-mentioned collaborations are driven by the application of EUCODIS technologies to generate novel products for our industrial partners and income for EUCODIS. In addition, we have a number of collaboration and service agreements in place which allow us to develop our own projects. The agreements related to the clinical development of our breast cancer project are of significant importance and include our:

- Service agreement with Nextpharma, UK, for the formulation and supply of clinical trial medication
- Production agreement with Merck, Germany, for the production of the active pharmaceutical ingredient (API) formestane
- Clinical development service agreement with the CRO Assign GmbH, Austria, coordinating the entire clinical development of formestane."

Academic collaborations

"Our core technologies were invented by some of the world's leading researchers at the Necker Faculté de Médecine. EUCODIS has considered it fundamentally important to maintain close ties with these academic groups. The reasons for this have been two-fold. First, this has given our researchers the opportunity to work on the premises of their academic colleagues, enabling us to develop the technologies rapidly for industrial applicability. Second, our continued participation in basic research allows us to maintain a leading edge over our competitors."

"Also, in developing our own projects, we consistently seek the advice and expertise of leading scientists in fields such as inflammation and bone diseases – an essential aspect to gain access to the tools and latest developments used in each specialty to maximize the chances of success of our projects."

Networks between industry and academia

"The partnership in the Competence Center for Applied Biocatalysis at Graz was and is of fundamental importance for the development of our business in industrial enzymes. The competence center is an entity comprising leading academic institutions in the field of biocatalysis and leading companies in the field (e.g., BASF, CIBA, DSM, Henkel, Sandoz, VTU). The competence center is financed by government funds and partner companies. Companies and academic groups either work on strategic projects (precompetitive area) to maintain and expand their scientific expertise and ensure rapid applicability for industry or they focus on bilateral projects for commercialization by industrial partners."

"This partnership allowed us to rapidly understand the market dynamics and the needs of the industry, has given us access to recent developments in biocatalysis, and has provided us with a stage to profile ourselves within this market. The Competence Center and EUCODIS form a strategic alliance to offer services to potential industrial partners. Several projects have successfully been conducted, and novel enzymes emanating from this collaboration are currently being developed."

Question 5 – What are the key challenges and lessons learned when it comes to financing and looking for alliances?

▶ "One of the main challenges is that, in every stage of your company's development, you need key players both inside and outside the company who fully support you – even when the going gets tough. At EUCODIS, we have been lucky to rely on highly experienced employees from the very day the company was founded, most of whom are still with us today more than four years later. All of our team have been instrumental in building up our network, both scientifically and financially speaking. This positive spirit and the enormous efforts of our team as well as our global network have been major contributors to our overall success."

"A major challenge in Austria is the difficulty of finding a follow-up investor after the first financing round and once the first public monies from government organizations have dried up. Being something of a hub between Central and Eastern Europe, Austria does provide excellent seed financing programs for start-up companies. However, Austria is not yet fully visible on the map of international investors, and the investment capital available in Austria does not match the financial requirements of the preclinical and clinical development of novel therapeutic approaches. This is an important aspect, because a number of Austrian companies are writing amazing success stories, being in an advanced development stage or even entering the market, such as Intercell, Nabriva, or Marinomed Biotechnology."

"Another key challenge is certainly the short-term outlook of most venture capitalists, many of whom follow the trends of their trade and its cyclic changes. Thus, the pendulum might swing from looking for investment opportunities in new technologies to exciting novel products and back again to new technologies. The time needed for bringing a promising research project into an advanced stage that carries a high value is often in sharp contrast to the rather short-time exit strategies of some investment funds."

"As company founder and CEO, you are expected to be an all-in-one genius who is able to use the language of the investor to make the value of his company understood. At the same time, a CEO has to be able to constantly motivate his people, particularly in strained times. Your staff is the backbone of your company that gives you strength and support at any time. Very often one ends up in a sandwich situation, and it takes a lot of time, energy, and patience to synergize the expectations of the market and those of your own company."

"One of the key learnings has been to see how important it is to rely on a close-knit international network of leading experts for individual projects to ensure optimal project design and research performance. You will also profit from this network in critical times and when you need urgent support. Also, entrepreneurs should start searching for a new investor as soon as one financing round has been closed. Failing to do so in time may leave the company in a hard spot later on, e.g., as a result of an unexpected economic downturn. In view of the tense situation on the global financial markets, we were lucky to have a US investor with deep pockets step in, who we had found through our excellent network contacts. Building these networks as early as possible is crucial for the later success or failure of a biotech company. When taking a big international investor on board, you also realize what the terms 'risk' and 'entrepreneurship' mean in the US or Asia and how differently they are still interpreted in Europe."

"Another key learning is related to our dual business model, being engaged in white and red biotechnology at the same time, and how differently these areas are perceived by investors. We rapidly found investors for our white biotech branch, EUCODIS Bioscience, focusing on enzyme production. White biotech projects carry a comparatively low risk, and the time horizon is much shorter. However, you do not have the prospect of reaching a value increase of more than 30% as in red biotech. Overall, we have had the luck – and foresight – to have been working in this area for several years now and are therefore ahead not only of the current 'hype' about white biotech but also of other companies only now entering this field. In the past years, we have gathered valuable experience from the services we supply to our business partners."

"Things are quite different in red biotech. It has much more appeal about it, nurturing excitement and enthusiasm, but you need a 'big shot' to convince an investor that he should take this higher risk and invest. You never enjoy quiet times in this area. From a financial perspective, you will have to struggle hard to manage a Phase III trial without a partner."

Question 6 – What will be the next steps for EUCODIS?

▶ "With regard to EUCODIS Bioscience and our white biotechnology projects, our short-term goal is to increase the revenue streams from collaborations, especially in the field of industrial enzymes. Looking further ahead, we expect to attain profitability from marketed proteins, developed by us either in-house or in collaboration with partners. In this respect, we have chosen a rather unique financing approach and will soon be able to offer real products. We will then see which path to follow – either pursue a strategic alliance, opt for organic growth, or choose a stand-alone solution."

"In the field of red biotechnology, our major target is to enter Phase II clinical trials with our lead project in breast cancer and to build up a product portfolio based on our specific platform technology. We aim to carry on with this very exciting project into Phase III and will then start our search for a global strategic partner. The plan is to conclude the Phase III trial successfully at the end of 2010. By then, we hope to have been able to build up the highest possible value chain. During the Phase III study we will very likely start searching for a partner. In our pipeline, we focus on three therapeutic indications, so that the search for a partner will primarily depend on the sales and distribution channels of these companies so we can achieve the best possible fit as well as long and sustainable results after the launch of the products."

"Overall, we certainly profited from our technology, which is widely applicable in different strategic business areas, ranging from enzymes to pharmaceutical drug development. This technology was definitely the basis for our value chain creation, and it might also enable us to buy in a new product candidate that fits this technology. Our technology platform enabled us to build up two different strategic business areas that appeal to different types of investors and their individual risk-taking capabilities."

4.4
Intercell AG

Interview with Dr. Werner Lanthaler, CFO — 11 March 2008

Before joining Intercell AG (www.intercell.com) as CFO, Werner Lanthaler was Head of Marketing and Communications of the Federation of Austrian Industry and, prior to that, senior management consultant at McKinsey & Company International. His academic accomplishments include a doctorate from the Vienna University of Economics and Business Administration and postgraduate degrees from Harvard University. He has considerable experience working in the labor and capital markets of the United States, South America, and Europe. Werner Lanthaler also is the author and co-author of a wide range of books and articles. Werner Lanthaler has been with Intercell since 2000.

Question 1 – When and how was the biotech start-up founded?

▶ "Intercell AG was founded in 1998 as a university spin-off from the Campus Vienna Biocenter. Focusing on developing vaccines for the treatment and prevention of infectious diseases, our company is working to leverage its proprietary technologies to develop novel vaccines, accelerate the time-to-market of products in late-stage clinical trials, and enter strategic partnerships to advance products in areas of high unmet medical need."

"The foundation of Intercell was driven by one vision: fighting infectious diseases and malignant tumors with the aid of a new generation of vaccines. This vision has come to fruition due to the scientific expertise of the company's founders Max Birnstiel, Alexander von Gabain, Michael Buschle, Walter Schmidt, and Aaron E. Hirsh. Alexander von Gabain, one of Intercell's co-founders, acted as the company's CEO until 2005. Previously, he had been Head of the Department of Microbiology and Genetics at the renowned Research Institute of Molecular Pathology (IMP) at Campus Vienna Biocenter, a basic research facility devoted to understanding the biology of disease development. Since the fall of 2005, Alexander von Gabain has served as CSO of the company."

"Today, our more than 250 highly qualified associates from 16 different nations put their expertise to work at our two Intercell sites in Austria and Scotland. The company headquarters and the research and development center are located in Vienna, Austria. Production, quality assurance, and process development are situated in Livingston, Scotland."

Question 2 – How has your company developed since its foundation and what products are you currently working on?

▶ "Our company's leading product, a prophylactic vaccine against Japanese encephalitis (JE), successfully concluded pivotal clinical Phase III trials in 2006. In the European Union, the vaccine was granted marketing authorization as an orphan drug, meaning that the vaccine can be sold exclusively for ten years. In the United States, a Biological License Application was submitted to the Food and Drug Administration in December 2007."

"With over three billion people living in endemic areas, JE, a mosquito-borne flaviviral infection, is the leading cause of childhood and viral encephalitis in Asia. The JE virus remains virulent in this region and has recently spread to countries not previously affected."

"Novartis will market and distribute Intercell's JE vaccine in the United States, Europe, and markets in Asia and Latin America. CSL is Intercell's marketing and distribution partner in Australia, and Biological E. will produce, market, and distribute Intercell's vaccine in certain Asian countries."

"Our company's broad development pipeline further includes a *Pseudomonas* vaccine in Phase II clinical development, a therapeutic vaccine for hepatitis C in Phase II as well as partnered vaccines for *Staphylococcus aureus* in Phase II and tuberculosis in Phase I. In addition, we have built up a pipeline of preclinical vaccine candidates, including a *Streptococcus pneumoniae* vaccine, a Group A *Streptococcus* vaccine, a Group B *Streptococcus* vaccine, a vaccine against traveler's diarrhea, and a bacterial vaccine with undisclosed indication (Table 4.2)."

"Further clinical studies will most likely include vaccine formulations with IC31® as a significantly more potent adjuvant and will be conducted under a co-development arrangement with Novartis."

"We are one of only a few companies worldwide to boast very short market lead times. From the outset, one of our primary strengths has been the combination of cutting-edge science with excellent product development, financing, and commercialization

Table 4.2 Intercell's product pipeline, January 2008 (www.intercell.com).

			Status	Expected milestones	Partner
Clinical vaccines	Japanese encephalitis	Prophylactic	Phase III	US/EU market approvals in 2008	Novartis (M&D)
	Pseudomonas	Prophylactic	Phase II	Phase II/III start 2008	In-house
	Hepatitis C virus	Therapeutic	Phase II	Phase II final data early 2008	Novartis
	Staphylococcus aureus	Prophylactic	Phase II	Phase III start 2009	Merck
	Tuberculosis	Prophylactic	Phase I	Phase II start 2008	SSI
	IC31® flu	Prophylactic	Phase I	Phase I data early 2008	Novartis
Preclinical vaccines	Pneumococcus	Prophylactic	Preclinical	Phase I start 2008	In-house
	Bacterial vaccine[a]	Prophylactic	Preclinical	Phase I start 2008	Sanofi Pasteur
	Group A Streptococcus	Prophylactic	Preclinical	Phase I start 2008/09	Merck
	Travelers' diarrhea	Cross-protective prophylactic	Preclinical	Phase I start 2008/09	In-house
	Group B Streptococcus	Prophylactic	Preclinical	Phase I start 2008/09	In-house
Antibodies (ABs)	Staphylococcus aureus	ABs in infected patients	Preclinical	Phase I start	Merck
	Pneumococcus	ABs in the elderly	Preclinical	Phase I start	Kirin
	Group A Streptococcus	ABs in infected patients	Preclinical	Phase I start	Merck
	Group B Streptococcus	ABs in premature newborns	Preclinical	Phase I start	In-house

[a]Undisclosed indication.

capabilities. Additional strengths have been our strong focus on value-added biotech products, our top-notch intellectual property management, and the availability of adequate liquidity reserves."

Question 3 – How has your company tackled the financing challenge?

▶ "We raised over €150 million through three separate fundraising rounds among top international investors, an atypical silent partnership, and an IPO (Figure 4.4)."

"Thus, in 1998, 2000, and 2003, we were able to raise a total of €75.7 million, the lion share stemming from international investors. In addition, an atypical silent partnership contributed €25 million to the first financing round."

"Austrian public funds, such as the AWS, WWFF, and FFF, played a crucial role at the start, and continued public support reflects both the importance of Intercell to Vienna as a business location and to society as a whole. Today, however, public money represents only some 8% of our company's capital base."

"Still, every Euro counts, and it counts double if you don't have to pay it back. Whereas venture capitalists expect their monies to yield high returns, public authorities are looking for different kinds of payback. Since its inception, Intercell has created more than 200 jobs, so I do believe that we have more than met the expectations of our public sector investors."

"Following these venture capital financing rounds, the next step towards strengthening our company's equity capital base was Intercell's IPO at the Vienna Stock Exchange in February 2005, opening the company to a wide base of investors. An issuing volume of €52.2 million resulted in net proceeds of €46 million, invested primarily into the development and commercialization of product candidates and the continued advancement of cutting-edge technologies."

"Intercell has a total of 45 521 707 issued shares of common stock, including 385 889 shares of treasury stock. Our major shareholders are Temasek Holdings (8.7%) and Novartis Vaccines & Diagnostics (15.9%). Currently, 0.8% of issued shares are held by Intercell as treasury stock, and 2.0% are held by our management. The current free float is 72.6% (Figures 4.5 and 4.6)."

Year	Financing rounds (€ million)
1998-1999 VC-1	5
1999-2000 K&W	25
2000 VC-2	27
2003 VC-3	43.5
2005 IPO	52.2
Total	152.7

Figure 4.4 Intercell's equity capital financing (€ million; Lindinger, 2005).

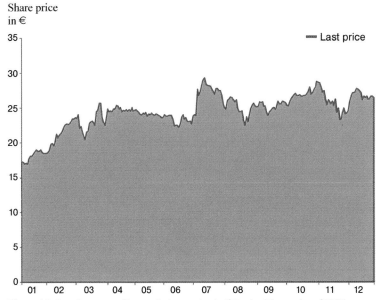

Figure 4.5 Development of Intercell share price in € in the 12 months of 2007.

"We have been generating revenues from public subsidies as well as collaborations and licensing agreements since our first operative year. The year 2001 saw the first revenues from advance payments from technology cooperations. In 2004, revenues from cooperations amounted to €3.4 million, or almost 75% of the year's total revenue of €4.6 million. In 2007, Intercell turned profitable earlier than planned and even before we brought our first product to market. Driven by a strong increase in

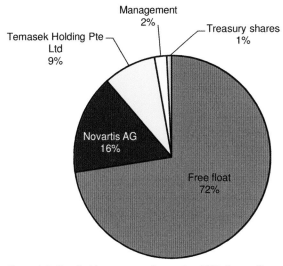

Figure 4.6 Shareholder structure, September 2007 (Intercell).

revenues from collaborations and licensing, we achieved our first positive full year result with a net profit of €5.0 million (compared to a net loss of €16.1 million in 2006). Given the significantly lowered risk profile of the company through a diversified portfolio of product candidates, world-class partnerships, and a very strong cash position, we have built a solid basis for innovation and aggressive growth."

"In terms of outside capital, some biotech companies consider credit financing a viable option. At Intercell, this financing tool has not played a role, reflecting the hesitation of banks to invest into biotech start-ups in view of our lack of collaterals. Neither did we rely on business angels, incubators, corporate venture capital, or mezzanine capital. Overall, therefore, Intercell's capital endowment was based on three venture capital financing rounds, contributions from an atypical silent partnership, the company's IPO, cooperation and licensing agreements, and federal funding, with venture capital playing the most important role."

"In the third financing round in 2003, the international consortium around the lead investor Global Life Sciences Ventures raised a total of €43.5 million in capital, with the world's largest life sciences investor, MPM Capital, contributing a substantial share of €13.5 million. Not only has this been the highest volume in venture capital ever attracted by an Austrian company, it was also the highest venture capital volume raised by the European biopharmaceutical industry between 2001 and 2003 (Figure 4.4)."

Question 4 – How important are cooperations?

▶ Financially speaking...

"The financial commitment of €100.5 million from private venture capital investors and an atypical silent partnership was based on two important prerequisites. First, from the beginning of the funding processes, Intercell's focus was on becoming one of the best global biotech companies. Second, we were well aware that a biotech company of this size would be unable to be financed solely with Austrian venture capital."

"To successfully complete a venture capital financing round, it is essential to get an international lead investor on board. We knew that, even internationally speaking, there were only some 30 biotech venture capital funds fulfilling this requirement in terms of financing volume, the specific know-how involved, and the networks that would be needed, and the company was able to win over some of the most internationally renowned lead investors, such as Techno Venture Management, Apax Partners, Nomura, and Global Life Science Ventures. The most decisive investment criteria certainly included marketability of the product, Intercell's highly advanced technology platform, the progress in clinical development, the management capabilities, and the large number of strategic alliances already in place."

"We greatly profited from the value-added services offered by our venture capital companies, such as MPM Capital. Specific advantages include the venture capital company's in-depth knowledge of the specific problems confronting the biotech industry, their proactive development of business scenarios, and their keen awareness that investments in the biotech sector require patience and stamina. Additional benefits of cooperating with international investors include their expertise and experience in managing later financing rounds and IPOs. Their supervisory function was carried out through a seat on the supervisory board."

Scientifically speaking...

"Scientifically, cooperation agreements are immensely important – they are something of a life insurance for the company. We have a number of highly renowned strategic partners. Not only are they a source of added value – to have international market leaders, such as Merck and Sanofi–Aventis, specifically opt for Intercell's vaccine technology is equivalent to having one's technology validated by external experts."

"Therefore, in addition to the vaccines developed in-house, selected product candidates, technologies, and development programs are offered to third parties for commercial collaboration and co-development partnerships, usually on the basis of licensing arrangements. Currently, Intercell is involved in the development of vaccines in the partnerships listed in Table 4.3."

"In July 2007, we signed an important strategic partnership agreement with Novartis to accelerate innovation in vaccines development in infectious diseases. This partnership happened to be the first of its kind in the vaccines sector and provides both companies with a strong base for mutual value creation. An upfront total cash contribution of €270 million has enabled us to further expand resources behind our key value drivers and secures the company's ability to independently achieve sustained aggressive growth."

"Another major vaccine partnership was started with Merck & Co., Inc., to develop a prophylactic vaccine against Group A *Streptococcus* infections. Under the agreement, we licensed certain antigens identified by our Antigen Identification Program (AIP) which have shown promising profiles in preclinical vaccine models."

"In December 2007, Merck initiated a Phase II clinical trial for the *S. aureus* vaccine candidate. The study is designed to evaluate the efficacy and safety of a single dose of the candidate vaccine in patients undergoing elective surgery. The start of this Phase II trial triggers a milestone payment of US$4 million to us."

Table 4.3 Current co-development agreements, January 2008 (www.intercell.com).

Indication	Partner
Japanese encephalitis virus vaccine	Novartis (M&D)
Hepatitis C virus vaccine	Novartis
Staphylococcus aureus vaccine	Merck
Tuberculosis vaccine	SSI
IC31® flu vaccine	Novartis
Bacterial vaccine (undisclosed indication)	Sanofi Pasteur
Group A *Streptococcus* vaccine	Merck
Staphylococcus aureus antibodies	Merck
Pneumococcus antibodies	Kirin
Group A *Streptococcus* antibodies	Merck

"We also collaborate with Kirin to develop human monoclonal antibodies against severe infections caused by *Streptococcus pneumoniae*. Our novel IC31® adjuvant was endorsed by a new partnership with Wyeth Pharmaceuticals, who were granted a worldwide non-exclusive license option for the use of our adjuvant IC31® in various selected infectious disease vaccine programs. Upon successful development of these programs, we are entitled to receive up to US$77 million in option and milestone payments as well as royalties on future product net sales."

Question 5 – What are the key challenges and lessons learned when it comes to financing and looking for alliances?

▶ **Financially speaking**...

"The only way to build up a biotech business is to attract investors early on. After that, it is essential to continue to meet the needs of existing investors and to take new investors on board. In terms of an optimal shareholder structure, the goal is to have between three and five investors with equal holdings. This ensures the best spread of risk, adequate resources for future investments, and the ability to take quick decisions in fast-moving markets."

Scientifically speaking...

"A well balanced product portfolio is crucial for the success of a biotech company. By clearly defining product candidates and advancing them through clinical trials, we always aimed at developing a balanced portfolio of promising product candidates to generate a regular and sustainable flow of new products. With the acquisition of Pelias in January 2007, we were able to add another major component to our product portfolio, i.e., vaccines and antibodies against nosocomial infections."

"Patents are the backbone of biotech. Our commercial success depends heavily on our intellectual property. We act to preemptively protect all inventions and collaborative research of potential commercial value. To date, our patent portfolio includes about 40 active patent families, covering both product candidates and technologies. AIP technologies and products, including more than 1000 antigens from different pathogens, are covered by more than ten patent families."

"Another main effort involves in-depth analyses of freedom-to-operate issues associated with all of the company's pipeline products. Upon identifying any third-party patent rights that may affect the development and operation of our products and processes, we either proactively negotiate to acquire a license from the patent holder or instigate invalidation proceedings. This approach led to the acquisition of a number of worldwide exclusive licenses for the development and commercialization of two of the company's clinical product candidates, the hepatitis C virus vaccine and the Japanese encephalitis virus vaccine."

Question 6 – What will be the next steps for Intercell AG?

▶ "Intercell focuses on market opportunities in areas with high unmet medical needs. The progress in clinical development is based on a strong commitment to penetrate the target markets as quickly as possible."

"In this context, 2008 is going to be a crucial year in our history. Initial revenues generated by the JE vaccine are difficult to predict, because the product will first have to earn trust in the market. In this respect, our distribution cooperation with Novartis holds much promise. Initially, the strategy will focus on high-risk groups, not least because the manufacturing capacity available at our Scotland plant currently is limited to one million doses of vaccine. However, expansion of the production capacity is already being planned, and the overall market potential is estimated at €250–350 million. The predecessor vaccine will soon have disappeared from the market, and neither in Europe nor in the US is there a competitor product in sight."

"Also, the year 2008 appears to hold additional cooperation agreements in store for us, making us confident that new and significant partnerships will be entered into in the areas of antigen product candidates. The company will continue to pursue its strategy of high-value partnerships, i.e., it will team up with a prospective partner only if the cooperation will clearly have a multiplying – and not merely an additive – effect on the strategic development of the company. Considering that, particularly in the biotech industry, not every product candidate will ultimately turn into a successful product, our long-term strategy is to develop our portfolio into one with just the right mix of risk."

"We will continue to strengthen our position as a highly innovative biotech company in the vaccine and antibody markets. Our first products will soon be brought to market, and in close cooperation with its partners, it will develop new and better vaccines. Significant product developments will be financed with our own cash flow. Most importantly, we will not be a company manufacturing just any product – instead we will offer products that will make a real difference in combating some of the most rampant and severe infectious diseases worldwide."

4.5
onepharm Research & Development GmbH

Interview with Dr. Bernhard Küenburg, CEO — 1 April 2008

Bernhard Küenburg is the founder and CEO of onepharm Research & Development GmbH (www.onepharm.com), an innovative pharmaceutical company focusing on chemical, clinical, and formulation development of small-molecule drugs for the treatment of viral as well as chronic inflammatory diseases. Before founding onepharm, Bernhard Küenburg was Senior Vice President of Marketing, Sales, and Development at Siegfried, a leading custom manufacturer of small-molecule drug substances in Switzerland, where he was responsible for strategic collaboration contracts, customer contacts, and strategic pipeline development. Before that, he was General Manager at CarboGen Laboratories, where he managed the construction and operation of the CarboGen Neuland facility. Previously, he was department head at Sanochemia in Austria, responsible for the development and manufacturing of the Alzheimer drug galanthamin. Bernhard Küenburg holds a PhD in organic chemistry from Vienna University of Technology.

Professor Dr. Otto Doblhoff-Dier, COO and co-founder, was previously Vice President for global manufacturing at Aphton/Igeneon. Prior to this, he was professor for bioprocess engineering and headed construction and production at the GMP pilot plant of the Institute for Applied Microbiology at the University of Applied Life Sciences, Vienna. Additionally, he is an expert on biosafety and member of various international biosafety boards and working groups. He holds a PhD in biotechnology.

Question 1 – When and how was the biotech start-up founded?

▶ "onepharm GmbH was founded in July 2005. The idea for its foundation had been born in the fall of the previous year after intensive discussions with a close friend of mine, the scientist Professor Otto Doblhoff-Dier (Figure 4.7). We had always wanted to run a company together, but for some reason the time to pursue such an important step had not yet come. This changed in 2004. I wanted to stop commuting to and from Switzerland for private reasons, and at the same time Otto's former business environment with Igeneon had changed dramatically. From the time we first started to plan our business idea until approval of the seed financing round by the Austrian government organization responsible for the funding of start-up companies with public money, the Austrian AWS, only half a year passed. In June 2005, Otto and myself were among the finalists for the 'Best of Biotech' award, an established business plan competition carried out by the Austrian government. There were 33 applications from five different countries in Phase I and 17 applications from four countries in Phase II."

"We started the company with two projects we obtained from another Viennese biotech start-up, AVIR Green Hills Biotechnology, because these two molecules were a better strategic and scientific fit with our own core competency, chemical synthesis. At the same time, AVIR Green Hills wanted to strengthen their own focus on biological entities, specifically influenza vaccines. This certainly was an amazing strike of luck, enabling the rapid foundation of our company with early seed financing acknowledgment. Moreover, we were able to win AVIR Green Hills Biotechnology as co-founders of onepharm and to profit immensely from their experience as co-

Figure 4.7 Members of the onepharm team: Dr. Oliver Szolar, Irene Zinöcker, Prof. Dr. Otto Doblhoff-Dier, Dr. Bernhard Küenburg.

Figure 4.8 Scientist in microbiological research laboratory.

partners. Ever since its foundation in 2005, onepharm has been an independent biotech start-up company concentrating its efforts on the development of new chemical entities rather than on new technologies."

"On 1 September 2005, onepharm became operational as a company. At first, it was only the two of us, Oliver Szolar, responsible for analytics and today's CSO, and myself in the role of CEO. Otto Doblhoff-Dier still had some transition work to close with his former employer and therefore joined our team shortly afterwards. The team continued to grow, and by January 2006 the core team had been established. Today, about 25 employees bring their experience and know-how to bear, either in the laboratory or as part of onepharm's management (Figure 4.8). In August 2007, onepharm moved its operations to the University for Veterinary Medicine, enabling us to combine laboratory and project management under one roof and to profit from the excellent infrastructure of a state-of-the-art university."

Question 2 – How has your company developed since its foundation and what products are you currently working on?

▶ "We have a product portfolio of small molecules that allows us to directly enter into preclinical and clinical trials in major viral diseases of the respiratory tract, such as the influenza virus, as well as in chronic inflammatory diseases like periodontitis. In addition, research programs will provide new lead compounds in these diseases and open up new indications for onepharm's substance library."

"In the months following the foundation of onepharm, we received much advice and operative support not only from AWS but also from my doctoral adviser and co-founder, Professor Ulrich Jordis from the Institute of Applied Synthetic Chemistry, as well as from family and friends. Moreover, in all questions regarding patent securement and the establishment of contracts, we were happy to rely on the services and competencies of an experienced patent lawyer."

"In terms of project development, we faced our first major challenge in the summer of 2006, when we decided, with a heavy heart, to stop the development of our first lead compound Elivir, an intranasal substance for rapid eradication of viral rhinitis. By the time we took the decision, a lot of time, energy, and budget had already gone into this project. However, looking back, this was the right decision to take, considering that the product stability was unsatisfactory and the toxicity of the compound's degradation products intolerable. The alternative would have been to continue through Phase II and then stop the project with a failure of meeting the endpoints, which would certainly have tarnished our excellent relationships to our investors."

"Honestly speaking, we went through a difficult period at that time, trying even harder to identify alternative projects for clinical development with the funds still available. The complex molecule that would finally enable us to return to our successful path was OPM-3001, a natural product extracted from the liquorice root which has long been known to have medicinal properties. For us, its activity against influenza seemed to be most promising. In late 2006, we succeeded in signing an exclusive cooperation agreement with the Japanese company Minophagen, which is the sole owner of the preclinical and clinical data package of this substance. Within record time of only four months, a binding letter of intent was set up between Minophagen and us and another six months later an extensive license and supply contract was successfully closed. These important milestones were achieved between year end of 2006 and mid-2007 and marked a new upswing in the development of our company."

"Despite this spirit of optimism with OPM-3001 we continued to engage eagerly in the search for additional compounds that would fit into our current development program and core competencies to secure our long-term company prospects. After a number of setbacks along the way, our perseverance was finally crowned with success. The newly in-licensed molecules fit our current portfolio perfectly and are poised to fill the treatment gaps in chronic inflammatory indications, such as periodontitis."

Question 3 – How has your company tackled the financing challenge?

▶ "After successfully securing our seed financing by the Austrian AWS and the 'high-tech double equity' programs of the FFG, the first major financing round with private investors was closed in June 2006, followed by an atypical silent partnership we entered into at the end of the same year. One year later, in June 2007, there was a new financing round supported by private and institutional investors and a new atypical silent partnership, leading to a capital infusion of more than €2.5 million in 2006 and 2007. Based on these achievements, we were able to apply for new funds and, luckily, were successful again."

"From the beginning, support by Austrian grant organizations played an extremely important role at onepharm. We succeeded in securing a substantial 'Vienna Spot of Excellence' grant from the ZIT as well as several major grants from the FFG. We truly believe that with AWS, FFG, and ZIT as well as with the tax model of the 'silent partnership,' Austria is one of the best places in Europe to support the foundation and growth of a biotech start-up company."

Question 4 – How important are cooperations?

▶ **Scientifically speaking**. . .

"From my personal experience, cooperations play a key role from the moment a small research-oriented company is founded – they can make or break a young enterprise. You can think of onepharm as a 'virtual company.' Thus, far over 50% of our work is delivered in form of special analytics, toxicity studies, stability testing, or clinical trials with established service providers, such as contract research organizations, preclinical and clinical testing laboratories, and other pharmaceutical companies located throughout Europe, such as Austria, Germany, Slovakia, the Czech Republic, or the Netherlands as well as in Japan, Russia, and the US. It is certainly a challenge to manage such cooperations over the distance, all the while proceeding in a timely manner. However, our highly motivated and professional team of preclinical and clinical project managers make this happen. Small companies will only be able to function properly if they gather together in research clusters with other specialists. And because budgets are limited, you want to select highly professional partners to work with, because even the minutest mistake may force you to repeat individual development steps, leading to delays and potentially having an enormous impact on the overall well-being of your project."

"Also, we have established a number of cooperations with international universities such as the Universities of Birmingham, Frankfurt, Regensburg, Denver, and Vienna. We maintain permanent working relationships with almost a dozen universities and even more external partners, and this network is vital to our company success. Last but not least, we always seek early advice from regulatory authorities to guarantee that our development programs are focused and target-oriented."

Financially speaking. . .

"One of the best ways to identify suitable investors is to participate in networking conferences such as those staged by BIO (e.g., BioEquity Europe, Bio-Europe Spring®), bringing together international decision-makers from all sectors of the biotechnology industry and providing life science companies with high-caliber partnering opportunities. At onepharm, we maintain continuous contacts with a number of private equity and venture capital investor companies that we update constantly on the latest developments of our products. Moreover, we will definitely approach new investors for new, larger financing rounds and introduce to them the current status of our development projects."

Question 5 – What are the key challenges and lessons learned when it comes to financing and looking for alliances?

▶ "One of the key learnings is that there is money available for financing biotech start-ups such as onepharm, however fiercely fought over the funds may be. At the same time, reaching a financing agreement with an investor takes time, usually between six and nine months."

"You have to find a way of bridging this time gap, because there is nothing you can do to speed up the process from making meeting arrangements and preparing for and holding the meetings to contract negotiations until final closure of the new financing round."

"Also, large investors may be engaged in several concurrent projects in different development phases – from Series A financing rounds to exit preparations, and you may see their priorities shift practically over night as a result of unforeseen events in one of their investment companies. In such cases, your own company may be put on hold – this may not actually have anything do to with your own projects but will affect you heavily and sometimes unexpectedly. This is why we will start our first talks with venture capitalists for the next major bridging financing round within the next few months. This lead time is necessary for an adequate and profound budget planning process well in advance of each new financing round."

"Another fascinating challenge derives from the fact that in the biotech area you have to work globally from the first moments of the company foundation onwards and that you therefore also need a certain infrastructure such as professional data management and quality assurance systems, a strategic company positioning statement with a USP, and a company website to keep pace with the big players in this field. This infrastructure has to be established with very limited human and financial resources. The good news about building up such professional systems is that you also continuously build up and move along your value chain. In the early phases of a company foundation, you depend heavily on the good relationships with your relatives, family, and friends, who in our own case provided some of their professional services even free of charge. We are still immensely grateful for all the 'start-up services' we profited from and still keep regular contacts with our early company supporters through invitations to our own networking events."

"Another key success factor for onepharm has been our strong team that has stood behind Otto and myself at all times, even when there were not too many good news to celebrate. The formation of a highly qualified, homogenous team of scientists and managers which not only cooperates well at a human and personal level but which is also willing to bear unpleasant situations is fundamental to the success of any high-risk business. As CEO of a start-up, you always have to be prepared to handle borderline situations, while at the same time continuing to motivate your team to work towards achieving our targets together rather than giving up."

Question 6 – What will be the next steps for onepharm GmbH?

▶ "We plan to close the Series A financing round of €8–10 million by the end of this year or early 2009. By then, we must have generated new and promising data points enabling us to convince our investors of our future growth and success. Moreover, we plan to move ahead of the competition with one to two new patent submissions that will further push our own value chain."

"Therefore, we have to work hard and with full speed in the upcoming months to achieve our new project milestones to secure for our business the next necessary financing round. This is an ambitious goal if you consider how much time it takes to initiate and successfully close talks with future investors and how unpredictable science can be."

Between conducting these interviews and going to press, some of the companies portrayed above had witnessed rather remarkable developments - some positive, others less so. These late-breaking news are summarized in the Epilogue.

5
Summary and Conclusions

5.1
The Challenge: When Science Meets Business

Entrepreneurs and venture capitalists inhabit two complex ecosystems, each governed by numerous specific rules and interactions which at times can appear rather opaque to 'outsiders.' It certainly is a challenging task to reconcile the interests of entrepreneurs and investors. Both parties come from 'different worlds' and need to first find a way to understand and talk in the language of their potential partner. Among investors, too, interests can be quite diverging. Unless this is taken into consideration before (strategic) alliances are established, a young company start-up may be doomed to failure right from the start.

Networking opportunities can help increase mutual understanding and appreciation between entrepreneurs and their investors and increase the likelihood of a win–win situation for all parties involved. Each development phase – with its different types of investors – must be regarded as a professional partnership for a limited period of time only. The players must be aware that it is of utmost importance to build trust – and that this takes time.

When it comes to entrepreneurship, it is essential for young scientists to be stimulated early on in their academic careers to develop their entrepreneurial thinking, e.g., through lectures on entrepreneurship and on what it takes to set-up a biotech spin-off from university. Again, some countries are ahead of others in this respect, but knowledge today is spreading quickly between universities in different countries, and most countries have already succeeded in setting up special institutions, such as bio-incubator centers and tech transfer offices supporting the creation of start-ups.

When it comes to financing options, it seems that it will hardly be possible to develop a biotech start-up without venture capital, even though, of course, there will always be exceptions to prove the rule. Most European countries are still struggling with obtaining follow-on funds after government funds have been used up and before venture capital steps in, leaving a financing gap for most biotech companies which, in most cases, cannot be overcome, no matter how creative the financing approaches taken.

Biotech Funding Trends: Insights from Entrepreneurs and Investors.
Alexandra Carina Gruber
Copyright © 2009 WILEY-VCH Verlag GmbH & Co. KGaA, Weinheim
ISBN: 978-3-527-32435-4

One of the key elements for improving the current situation is to educate the general public about the benefits of private equity and venture capital and to support efforts to create a more 'private-equity friendly' climate in Europe, following the example of the US.

Both investment and research follow a cyclic behavior. Investors often follow their peers and the general trends prevailing at the time, a behavior that may be regarded as opportunistic rather than strategic. Venture capitalists with some scientific background, who can rely on their own judgment and do not have to follow external consultants and general trends, are likely to stand greater chances of being successful, particularly in times when the economic situation and the private equity climate are not all that promising. The willingness to take an adequate amount of risk in insecure situations may well pay off – not only for the entrepreneur, but also for the investor.

5.2
Checklists

Based on the interview results obtained and the understanding of the major factors for successful funding and alliances between entrepreneur and investor, some practical guidelines were derived from the interviews on how to foster mutual understanding and on how to increase the odds of establishing a win–win situation for all parties involved in the development and funding of biotech start-ups.

5.2.1
Checklists for the Biotech Entrepreneur

5.2.1.1 What Makes a Perfect Entrepreneur?
For the venture capitalist, it is most important to identify company founders with comprehensive leadership and interpersonal skills which demonstrate that the entrepreneur can share responsibilities, work in teams, and maintain a professional partnership with a venture capitalist. Company founders will need to share responsibilities and duties later on as the company grows and even be willing to step back and hand certain responsibilities to outside personalities. Second, the entrepreneur should be an experienced individual, ideally with a track record in the industry and capable of taking the lead in technologic and strategic decisions. Bringing an entrepreneurial spirit to the job is also regarded as desirable, considering the multitude of challenges that need to be resolved in the first years of a start-up company. The third major strength required of an entrepreneur is a profound scientific and operational knowledge of how a company can best be kept up and running. For the group of other investors, additional characteristics were a strong belief in the company and its uniqueness as well as a clear vision and commitment of the entrepreneur.

5.2.1.2 What is the Most Important Entry Criterion for an Investor?
Venture capitalists will evaluate two major aspects before they decide to invest in a biotech start-up. First, the scientific rationale has to be convincing, perhaps even

providing a new technology platform. Second, the management team has to be experienced and characterized by entrepreneurial thinking. Only if these two factors are fulfilled does the company stand a chance of surviving the years of hard work and endeavor. Moreover, not only do science and management have to be convincing, attractive, and appealing to the investor, they also have to make a perfect match. Other important criteria are credibility, a USP, and solid IP protection. Particularly the latter two factors will be responsible for sustained and long-term market opportunities for the start-up.

For the group of other investors, the uniqueness of the business model was equally important as the scientific rationale and the management team. This result suggests that for this group the 'market' factor is essential for the sustained success of the biotech start-up to clearly differentiate the products from its competitors or substitutes.

5.2.1.3 How Can an Entrepreneur Best Deal With his Dual Role?

The entrepreneur should be aware of the dual challenge that is imposed on him, namely that of being a scientist and a general manager at the same time. In the early days of a company foundation, the entrepreneur will lead the process of business plan preparation, determine the strategic outline, plan the budget, and acquire some more general management and leadership skills. However, he should later be willing to share responsibilities and tasks with an experienced CEO from outside. Different development stages call for different sets of skills to have the company value increase in the most effective and sustainable manner. Of course, one key success factor is identifying a suitable partner with complementary skills and experience. During this phase of role separation, it may be beneficial for some entrepreneurs to consider seeking support through coaching to find out more about their own strengths and weaknesses and to learn to let go.

5.2.1.4 How to Select the Right Fund and the Right Fund Size?

In early development phases, i.e., at a time when the entrepreneur can profit most from local support and counseling, a local or regional fund office is preferable. Also, a regional fund usually can provide local public relations and lobbying to attract global funds for the next financing round and usually has a wider experience than a purely national fund. In later development stages and as funding requirements increase, cooperation with a global fund should be sought.

In Europe, two additional factors make investments more difficult than in the US. First, there is an evident difference in cultures across Europe, even between neighboring countries, and language barriers can also make across-border business something of a challenge. Second, venture capital funds in Europe have to deal with different national regulations between member states, a problem that has now been taken up by the European Commission. A first step towards improving the current situation will be to eliminate the double taxation within the EU, the final goal being the free movement of capital between member states. Some interview partners added a critical remark about the prime focus of many funds, with the biotech industry often underrepresented because of the longer development horizons compared to other high-growth start-ups, such as IT or telecommunications.

5.2.1.5 Which Funding Mechanisms are Available to European Biotech Start-ups?

European biotech start-ups usually rest on three major capital sources, i.e., government grants, venture capital, and strategic alliances. In some countries, such as Austria and Germany, government grant and loan arrangements play an important part in seed financing. In other countries, such as Switzerland and the UK, there are nearly no public funds available, but there is a significant number of business angels and early-stage venture capitalists who usually step in early in the development process. Therefore, the optimal mix of financing forms appears to be country-specific.

The picture is more diverse and complex for the remaining financing forms. For the entrepreneur in Germany and Austria, silent partnerships are essential early-stage financing options. Another important financing form whose importance will continue to increase are business angels who, in the Anglo-American culture, step in rather early and frequently. They support the entrepreneur with their knowledge, experience, and network in a phase too early for the venture capitalist to invest. EU grants are seen as an important new source of additional funds, although the processes from submission to approval can take quite some time and require a lot of administrative work. Other financing forms, such as corporate venture capital, mezzanine capital, and debt financing, were found to be less important.

Thus, the funding of biotech companies rests on three major pillars and a certain number of additional, creative sources used and combined differently by various biotech companies, demonstrating that entrepreneurs tend to find innovative ways of financing their expensive and exhausting R&D research.

5.2.1.6 What is the Primary Financing Form for Entrepreneurs?

When it comes to the primary financing form used by biotech start-ups, the picture drawn by entrepreneurs was rather diverse and differed significantly between interviews. For example, two entrepreneurs reported to successfully rely on private equity as their major source of financing, whereas two others had had the opportunity to partner with big pharmaceutical companies early on. Entrepreneurs are usually aware of the fact that it is essential, though very difficult, to obtain venture capital in early-stage financing and are therefore forced very early during their development to seek for alternative solutions to continue with their research. In addition, many entrepreneurs seem to be aware that, if they succeed in another financing round with an important partner, new additional capital investors are more likely to follow. Yet, they know that the down side of an early strategic alliance with 'big pharma' is that it might deter additional investors or pharmaceutical partners.

5.2.1.7 What are the Major Advantages and Disadvantages of Venture Capital?

The benefits of venture capital, also referred to as 'smart money' because it secures the entrepreneur experienced partners and strong networks, are that it is risk capital made available to a high-risk industry, i.e., the life sciences, without the need for personal securities by the company founder. For this reason, it is the most important financing source for companies that are in an early project development stage, which is too risky for major pharmaceutical companies to get involved. Venture capital is true 'private equity' from professional partners with profound experience from previous projects – or 'corporate memory', and good networks, enabling them to give meaningful strategic recommendations. The disadvantages of venture capital are that it 'dilutes' the entrepreneur and forces him to give up certain of his rights and that it is sometimes provided with fast outcomes in mind. Also, venture capital funds in Europe are still limited, with only a minute fraction of applicants being granted funding.

5.2.1.8 What are the Major Advantages and Disadvantages of Government Grants?

The benefits of public 'soft' money are that it is available for investment into a business idea very early, it does not dilute the entrepreneur, and it is related to an early evaluation of and support for the new biotech start-up. Its drawbacks are the bureaucratic efforts involved in applying for public monies, the time it takes until the funds are finally granted, and the influence government organizations some-times try to exert on the management of the start-ups they support.

From the entrepreneur's perspective, government grants and loans are highly important as preseed and seed financing tools. Today, government organizations are better prepared and equipped than in the past to support future entrepreneurs in determining the most suitable strategy for their companies, not only in terms of developing a business plan, but also in helping the founder translate this plan into reality and in accompanying and supporting the start-up to reach its milestones.

Generally, the positive effects of government grants outweigh any possible draw-backs. Nevertheless, loans may have a negative impact on the balance sheet by diluting the company value. The evaluation process of government organizations is sometimes seen in a critical light because too many start-ups receive preseed and seed financing on a rather indiscriminate basis, even though many of them will not be able to raise follow-on capital. Thus, a more stringent process on the part of public bodies in supporting start-up companies would be beneficial for the funding process as a whole. Some hold that government organizations should not offer any public money to support the seed financing phase of young companies, because such 'donations' only offer short-term support without encouraging entrepreneurs to develop a long-term commercial perspective on where to take the company.

5.2.1.9 What are the Major Advantages and Disadvantages of Strategic Alliances?

The benefits of partnering with a major pharmaceutical company are that it considerably increases the value of the company, spreads risk, has a strong signaling effect towards the public and other investors, does not dilute the entrepreneur, provides opportunities for scientific cooperation, and opens up access to marketing and distribution channels.

On the downside, partnering with a big player too early may be tantamount to giving away much of the company's assets at a significant discount. Also, large companies tend to be less flexible when it comes to the fulfillment of milestones and accepting possible delays. In addition, early alliances may lead to the start-up becoming less attractive for alliances with other pharmaceutical companies. Finally, entrepreneurs should be aware that processes in large corporations have long lead times, potentially delaying important decisions the start-up may need to take.

5.2.1.10 Are There Any Creative, New Financing Approaches?

The biotech financing world appears to be colorful and varied. Thus, research collaboration alliances and contract research are of high importance. Another creative financing approach mentioned was related to a mix of two different types of investors working together on one company project, e.g., a venture capitalist with a business angel or a bio-incubator with a corporate venture capitalist. This shows that individual investors are willing or seek to share competencies and split capital costs to finance the expensive R&D efforts of companies they believe in. Other creative financing models mentioned were project financing, also referred to as the 'virtual biotech model,' new stock markets in Europe, and high net worth individuals.

5.2.1.11 What are the Determinants of Success When Selecting a Biotech Research Area?

Three basic factors identified by the interview partners determine whether a biotech research project stands chances of being successful, i.e., the market potential, future biotech trends, and the depth of its portfolio. In terms of market potential, the market size and the innovative potential of the drug candidate are important. Generally, any area with an unmet medical need can be an attractive investment target. Another essential aspect is the treatment duration, with indications requiring longer-term treatment being more attractive than short-term therapies. Here, cancer treatments are still high on the list of attractive drug candidates, and many experts are convinced that cancer will soon be a chronic disease rather than a fatal diagnosis. At the same time, interview partners stated that there was hardly a research area more risky and challenging than oncology. Finally, adequate intellectual property protection needs to be put in place, and the business model must be sales-driven.

Promising indications also include those that attend to upcoming or future healthcare trends and will gain importance in the future, e.g., the prevention of disease through vaccines against cancer, the prediction of disease, and personalized medicine. One of the slogans of the future will be 'from sickness to wellness' (Burrill, 2007). Also, it will be important to attend to the needs of an aging population and to develop means to foster health management in the elderly. Particularly from the vantage point of venture capital, the portfolio should always be as diversified and balanced as possible to spread and reduce the risk across the different biotech companies.

5.2.1.12 How Can Entrepreneurs Get in Contact with Venture Capitalists?

Highly efficient ways to meet venture capital companies include bio-network conferences organized on a regular basis or the participation in business plan competitions. Another possibility to indirectly establish contacts with venture capitalists is through commercial intermediaries, such as incubators, business angels, tech transfer offices, and government funding organizations.

5.2.1.13 What is the Preferred Deal Structure for Venture Capitalists?

There was common agreement among almost all interview partners on what the most favorable terms for the investor were, regardless of the development stage of the company. Thus, the two most popular arrangements with entrepreneurs were convertible preferred shares and a liquidation preference. Convertible preferred shares enjoy certain preferential rights, such as an anti-dilution provision, giving venture capitalists the right of first refusal over any future issue of equity to outside investors. In such an agreement, the anti-dilution provision adjusts the number of shares (or the percentage of the company) held by the holders of the preferred shares upwards if the firm subsequently undertakes a financing round at a lower valuation than the one at which the preferred investors purchased the shares. A liquidation preference determines the order in which different security holders are paid in the event of a liquidation. It ensures preference over common stock with respect to any dividends or payments in association with the liquidation of the company (Amis and Stevenson, 2001; Lerner et al., 2005; Metrick, 2007).

5.2.1.14 How Should the Board of Directors be Structured?

For the entrepreneur, the mix of expertise was the most important attribute for structuring the Board of Directors. Ranking second, third, and fourth were small size, independence, and active contribution. According to the venture capitalist, the board should be structured according to knowledge and influence in society, and the amount of influence should not necessarily be determined only by the size of the investor's contribution. Some interviewees stated that different development stages may call for different structures of the Board of Directors. For example, in the time period preceding an IPO, it may make sense for the venture capitalist to step back and leave the floor to experienced industry experts.

5.2.1.15 Why do Many Biotech Start-ups Fail?

For both entrepreneurs and venture capitalists, the scientific risk and mismanagement were mentioned as equally important reasons for failure. The third reason given was the lack of a partnership with suitable investors, leading to follow-up financing not being available, underinvestment, the inability to enter partnerships, and a lack of access to financial markets. One investor considered insufficient market analyses and another one-product strategies the most likely primary causes of failure.

5.2.1.16 What Countries Enjoy a Favorable Biotech Start-up Climate?

The majority of interview partners named the UK as having an advantageous start-up climate. This is due to its highly developed private equity culture, its cultural proximity to the private equity pioneer US, the availability of early venture capital funding from university-based funds, angel networks, the country's entrepreneurial culture, the strong pharmaceutical clustering, and the target-oriented approach to research with a high density of research biologists in the country. The

German-speaking countries also got very positive feedback from the interviewees. The most prominent arguments in favor of Switzerland were access to large sources of capital, strong biotech, pharma, and research clusters, the high number of business angels, 'smart' universities, the longer private equity and venture capital culture and the Anglo-American influence, and the entrepreneurial culture prevailing in Switzerland. The most important argument in favor of Austria was the good public seed financing programs, even though some companies are left without follow-up financing after the first round. Also, Austria still has to develop a strong entrepreneurial spirit and private-equity culture. The most important arguments in favor of Germany were the high level of scientific expertise and the availability of funding. Other countries with favorable biotech start-up climates mentioned were Ireland and Scandinavia.

5.2.2
Checklist for the Investor

5.2.2.1 What Makes a Perfect Venture Capitalist?

For the entrepreneur, an important factor mentioned was to be able make use of and profit from the regional or global network of the venture capitalist. Other factors ranking equally high were capital (access to financial markets), knowledge (a proper understanding of the underlying innovation), experience (a proven track record in the industry), and a long-term perspective (patience and an ability to anticipate the challenges ahead). Particularly this latter factor was considered key for a professional, successful, and sustainable partnership. It is this long-term perspective that the entrepreneur has to be able to count on, particularly in an area as time-consuming and unpredictable as biotechnological research and development. Venture capitalists should therefore signal their willingness and commitment to provide long-term orientation and guidance to the biotech start-up, also in case of upcoming obstacles and hurdles that may at one point slow down the progress of the company.

Some investors also pointed out that strengths such as interpersonal or negotiating skills were definitely important, but that their significance was often not evident to the entrepreneur. Investors who feel that they have a particular strength in these areas should make entrepreneurs aware of these important capabilities and give examples of when these strengths may become an important prerequisite.

5.2.2.2 How Can an Investor Best Support an Entrepreneur in Dealing With His Dual Role of Scientist and CEO?

During the phase of role separation, the investor should be willing to support the entrepreneur and company founder by finding a partner that best fits the company and its team. The venture capitalist will be responsible for buying in additional experiences and skills and for setting up a professional management team. Different development stages call for different skill sets, and failure of any member of the management team to understand this may require tough personnel decisions to be taken and the management to be reshuffled. Every start-up and its underlying business plan is unique and will therefore require strategic decisions to be taken on a case-by-case basis.

5.2.2.3 What Services Should a Fund Provide to an Entrepreneur?

Venture capitalists and their funds must be large and potent enough to fulfill a number of key requirements to satisfy the needs of the entrepreneur and his biotech start-up. These include the capacity to invest in companies outside their home countries, funds big enough to take the start-up to the next financing round, networking between the venture capitalist and the entrepreneur, and local offices close to the biotech start-up and its research premises to provide support and advice.

5.2.2.4 What Valuation Techniques are Applied in Practice?

Valuation methods can be categorized into methods based on discounted cash flows, methods comparing the company in question with a similar company in the same business sector, the venture capital method, real options, and the Monte Carlo method (Frei, 2006). Even though valuating high-growth, high-uncertainty companies is a challenge, particularly if they are combined with accounting losses, discounted cash flow (DCF) remains a useful tool. Alternatives, such as the price-earnings multiples, generate extremely imprecise results (earnings are highly volatile), often cannot be used (when earnings are negative), and provide little insight into what drives the company's valuation. More important, these shorthand methods cannot account for the uniqueness of each company in a fast-changing environment. Another alternative, real options, is certainly promising, but current implementation techniques still require estimates of the long-term revenue growth rate, long-term volatility of revenue growth, and profit margins – the same requirements as for discounted cash flow (Koller et al., 2005).

The results show that in earlier phases, the valuation techniques considered acceptable by the interviewees are quite diverse, ranging from comparables, the venture capital method, valuation of the management, and gut feeling to discounted cash flow, real options, and even Monte Carlo analysis. In later development stages, there is a shift to fewer and more sophisticated techniques, i.e., DCF, Monte Carlo simulations, or real options.

5.2.2.5 What are the Differences Between Classic Biotech and Vaccines?

Regarding the possible differences in the business models between a classic biotech product, such as antibody development against cancer, and a vaccine, interviewees stated that business models should take into account possible differences in three areas, i.e., R&D, manufacturing, and market analysis. The following questions are therefore worth asking:

Research and Development

- Are there any government-induced incentives to start research on a particular vaccine, e.g., a vaccine against HIV?

- Can a partnership with a government or non-profit organization be entered into? Examples of organizations include the NIH, the military, and the Bill and Melinda Gates foundation.
- Does the clinical trial program focus on prevention or cure?
- How many subjects are needed for clinical trials? For example, for childhood vaccines, large sample sizes will be required.

Manufacturing

- Should an early strategic partnership for manufacturing collaboration purposes be entered into?
- What are the production capabilities?
- What are the manufacturing costs?

Market Analysis

- How large is the number of potential buyers?
- How many possible competitors are there?
- What is the market potential? Is the vaccine effective against an infection or disease prevailing in only some parts of the (developing) world, e.g., West Nile virus infection, or will the vaccine be able to be sold on a global level, e.g., a hepatitis vaccine?
- What pricing strategy will be pursued?
- How likely is the product to receive recommendation or reimbursement in the countries where it will be marketed?
- How can vaccination be provided to and financed in developing countries?
- How will the vaccine be commercialized? Will a distribution partner be required? Will it be possible to close tender agreements with governments?

5.2.2.6 What Factors to Consider with Regard to an Alliance Between a Venture Capitalist and a Corporate Venture Capitalist or Business Angel?

In general, the alliance between venture capitalists and corporate venture companies is regarded as a beneficial option, as long as big pharmaceutical companies do not claim any special rights, such as first right of refusal, right of utilization, or first option on the product. Right issues could clearly become a conflict of interest, because not only is a corporate venture capitalist accountable to his investor, i.e., the mother company, he is also obliged to sell the company at the best achievable market price. Therefore, syndicating is desirable provided that no special conditions are to be granted to the corporate venture fund. Moreover, as is the case for cooperations with business angels, corporate venture capitalists will also have to accept that, if they cannot finance a second round, they will be diluted and outvoted.

5.2.2.7 What is the Role of New Markets, such as London's Alternative Investment Market?

Among investors, there is still much insecurity when it comes to new markets, such as London's Alternative Investment Market (AIM). Of the 2900 companies currently listed on AIM, only a minority is derived from the biotech area. Even though UK-based biotech companies may find an opportunity to be listed on their home market very attractive, this 'home-country bonus' does not apply for companies outside the UK.

AIM does provide a market place for company presentation and gives early access to the capital market. Again, the final decision which stock market to choose will depend on the company's overall strategy, and the following questions should be addressed:

- How large is the market of my business?
- Where do possible investors look for investment opportunities?
- Where is the competition located and listed?
- How do I stay on the radar screen of the stock market?

5.3
Final Observations

Financing start-ups is definitely a challenge for any investor who decides to become part of the overall financing structure of a young company, regardless of his risk-taking capability. The central research question of this study was based on the general financing model matrix for innovative start-ups. Based on the financing matrix (Figure 2.2) and against the backdrop of the complexity of biotech development, the primary question was whether a biotech start-up went through the same development phases and was financed by the same types of investors as other innovative start-up companies, such as IT, industrial products, or services companies.

Because of the high risk and long development times involved in biotech, financing a biotech start-up appears to be more of a challenge than in other industries, such as IT, which usually enjoy shorter times to market and a clear exit vision from the first financing round onwards. As a result, in biotech the number of investors willing to step in after the seed financing stage (usually through government funding programs) is still limited, and biotech entrepreneurs often need to find creative approaches to bridge the seed and early-stage financing phases. Almost every interview partner in the entrepreneurs' group had taken an individual approach to financing their respective companies, looking not only which money was most easily available, but also which option produced the highest strategic and long-term value for the company, with options ranging from business angels, high net worth individuals, private-equity deals, and atypical silent partnerships to those not utilizing any venture capital – as the AVIR Green Hills Biotechnology case study showed. The farther the company moves along the value chain towards the late development stage, the lower the risk of investing in the company, the clearer the exit strategy – and the greater the willingness of the venture capitalist to invest. Also, the valuation of a late-stage company is better estimable than in the early stages. Many entrepreneurs appear to be aware that, if they succeed in another financing round with an important partner, additional investors are more likely to follow.

Most venture capitalists consider venture capital the single most important financing source for biotech companies. It is regarded as 'smart money,' supporting the entrepreneur not only with capital, but also with experienced partners and strong networks. However, venture capitalists also concede that this financing form has the major disadvantage of diluting the entrepreneur. For this reason, venture capitalists are well aware that company founders need to be incentivized to remain focused and motivated.

For all three groups of interviewees, the three most important financing forms for European biotech companies were venture capital, government grants, and strategic alliances. Especially in countries such as Austria and Germany, government grant and loan arrangements play an important part in the seed financing of biotech companies. In other countries, such as Switzerland or the UK, there are nearly no public funds available, but there is a significant number of business angels and early-stage venture capitalists to step in earlier in the development process. Therefore, the optimal mix of financing forms appears to be country-specific. These three major financing pillars are complemented by a limited number of additional creative approaches, such as project financing, research collaboration alliances, contract research, and new stock markets for small high-growth companies – all used and combined differently by each company.

Expectedly, some of the questions on interpersonal relationships and alliances in the field of biotechnology generated highly controversial answers, such as those on alliances between scientists and venture capitalists or venture capitalists and corporate venture capitalists. Obviously, every investor, in addition to wanting to make the biotech start-up a success, also has his own interests to defend against other investors.

Every phase in the drug development and financing process is clearly a 'people business', depending on individuals who are willing to build up mutual trust and rely on each other even in times of crisis. This is important, because the goals of the individual parties involved in helping a start-up get running might not always be congruent. For example, whereas big pharmaceutical companies may be reluctant to share their knowledge for competitive reasons, the small biotech start-up may want to share their know-how to expand and strengthen their network. Partners have to be carefully selected for the overall best interests of the biotech start-up. In case of diverging interests, it may be better to forgo such an alliance. Overall, whether the option selected will or will not succeed will depend on the individual business model, the company and its staff, and its investors.

The truth is neither black nor white. All interview partners were highly experienced professionals, open, aware of the many challenges a company may be faced with, and willing to share their thoughts and findings – brilliant people with brilliant ideas. Each interview was like a unique building block combining into a bigger whole, which – with the supporting method of hermeneutics – finally allowed general conclusions to be drawn. One of these general conclusions was that each biotech start-up is unique in the ways it conducts its research, operates, and finances itself. Also, with Europe being a culturally diverse economic area, this diversity and the different private equity backgrounds among European countries likewise have to be taken into account. As a result, each biotech start-up requires a highly personalized approach.

Biotech and funding are subject to global trends and cyclic developments, and biotech and funding trends do not always coincide. Investors tend to follow each other in their investment behavior. Because there is no way for young and small companies to influence these global trends and because only few investors behave in an acyclic manner and go against the general flow, it can become difficult for a biotech start-up to survive. Particularly in strained times, the further development of a fledgling company will depend on those brave and entrepreneurially oriented investors who, by drawing on their biotech know-how, are able to correctly and realistically evaluate the opportunities before them and who are prepared to take a risk even when the economic and financial environment is anything but favorable.

6
Outlook and Recommendations

From the insights obtained during the interviews, there is an increase in biotechnological research and development in many European countries, with a growing
number of university and corporate spin-offs being founded. Most of these start-ups
are still in an early development phase, but if they succeed in closing the gap between
seed government grants and early-stage venture capital, they will be able to further
develop their innovative product pipelines. The results of this survey suggest that the
climate for innovation and biotech companies in Europe is improving, thanks due to
increased know-how, the ever higher number of experts in this area, and more capital
from diverse sources being available.

From the private equity perspective, the challenge still is to obtain sufficient
resources to support these high-risk business opportunities. This requires know-
ledgeable personalities who understand the science behind the opportunity. Venture
capitalists prefer to invest in experienced entrepreneurs with a vision to transform a
business opportunity into a long-lasting value that represents an urgent and unmet
medical need, has a clear value proposition, has secured itself strong IP protection,
and whose proof of concept is already available.

None of the above is a walkover, neither for the entrepreneur nor for the investor. It
will become increasingly important to invest in lobbying and public relations
initiatives to help the general public understand how private equity investments
work and how they can benefit society. Most likely, venture capital spending will
increase in the coming years, and private equity companies will be willing to step in
earlier than in the past. Business angel networks likewise are going to gain ground,
and more high net worth individuals will engage in the financing of more biotech
start-ups. Overall, making more money available will be the key success driver for
innovation and R&D biotech research in Europe's near and mid-term future.

There are, of course, rather significant differences in private equity spending
among the European countries, with the UK, Switzerland, and the Nordic countries
being relatively advanced compared to other European countries, such as Austria and
some Eastern European countries, who are slowly catching up.

In many ways, the American way of financing biotech start-ups can be a learning
example for Europe. At the beginning of the value chain are highly professionally
organized tech transfer offices at universities that help young scientists with a great

Biotech Funding Trends: Insights from Entrepreneurs and Investors.
Alexandra Carina Gruber
Copyright © 2009 WILEY-VCH Verlag GmbH & Co. KGaA, Weinheim
ISBN: 978-3-527-32435-4

and innovative business idea found their own start-up. The primary financing source for many of these companies in this early phase are business angels, many of them former entrepreneurs themselves who find it fascinating to consult start-up companies once they have withdrawn from active business life. They are frequently organized in private networks, and they have the knowledge, the experience, the network, and the money – all factors that the entrepreneur at the beginning is highly dependent upon. After that, it is usually the venture capitalist who steps in, who is prepared to invest a significant amount of money and to take high risks for the further development of the company. Usually – and in contrast to European investors – they act quickly in order not to lose any major opportunity to a competitor. At the same time, they are keenly aware that they are acting in a high-risk industry, with only between one or two out of ten product ideas ever reaching the market. Next in line after the venture capitalist is the big pharmaceutical companies. Today, all major pharmaceutical companies, due to their own lack of innovative, new blockbuster drugs, usually get alerted to new, promising biotech start-up companies getting on their radar screen very early on, and, again, they will have to react quickly to be ahead of the competition.

The difference between the US and the European entrepreneurial cultures might be even more pronounced than that in the venture capital area. The US entrepreneurial culture is based on the freedom to fail. As a result, a track record of entrepreneurial failure is accepted, it is considered a way of gathering experience, and, for many entrepreneurs, results in a second chance. In Europe, this culture of failure is not yet highly developed. Instead, failure usually leads to stigmatization and the lack of a second chance.

Overall, however, Europe's venture capital and biotech scene is clearly following an upward trend, with an increasingly important network established between individual key players. Investors will have to cooperate more closely and build trust amongst each other. At the same time, they will need to continue to provide guidance to young entrepreneurs who take upon them the risk of founding a new company. Only once the interests among investors and between investors and entrepreneurs have been aligned will a win–win situation be able to develop, and this will most likely result in successful company developments. All parties involved need to understand from the very outset that they are entering a partnership for a limited time only and that, particularly in times of crises, there might be hard decisions to take. The overall goal of all parties should be to not pursue solely their personal interests but to achieve the highest value increase of the company and its products to write yet another European company success story. If this rule is followed, it will be much easier for all stakeholders to work together in taking the right decisions at the right time.

Epilogue

In the time between the interviews with the five Austrian biotech entrepreneurs and the time of going to press of this book, some of the companies portrayed have undergone rather significant changes – some highly encouraging, some rather disappointing. This epilogue, which pulls together the press releases published since the interviews were conducted, highlights once more that biotechnology takes place in a very dynamic and fast-changing environment.

Three of the five companies, i.e., AVIR Green Hills, Intercell, and onepharm, have been able to report significant progress in their research activities, collaboration agreements, or even acquisition endeavors. One company, EUCODIS, has not recently been in the news. However, one of the five companies, AUSTRIANOVA, has meanwhile had to file for bankruptcy due to liabilities amounting to more than €11 million.

AUSTRIANOVA Biomanufacturing AG

On 30 May 2008, insolvency proceedings were instituted for the assets of the Vienna-based biotechnology company AUSTRIANOVA Biomanufacturing AG, the holding company of AUSTRIANOVA Biotechnology GmbH, which had filed for bankruptcy on 9 May 2008.

The originally planned forced settlement of the subsidiary AUSTRIANOVA Biotechnology GmbH had become impossible to carry through due to a lack of financial resources. Therefore, the Vienna Commercial Court ordered closure of the company on 28 May 2008. According to the bankruptcy petition, the company's liabilities were estimated at €11 million and were primarily composed of grants and loans. 20 employees were hit by the closure.

According to the "Kreditschutzverband von 1870," Austria's largest association for the protection of creditors, the ensuing insolvency of the holding company, AUSTRIANOVA Biomanufacturing AG, is due to its having assumed liability for research loans granted to its now insolvent subsidiary. The liabilities of AUSTRIANOVA Biomanufacturing AG amounted to approximately €6 million, debtors said. Next to the Board of Directors, the company employed three associates.

Biotech Funding Trends: Insights from Entrepreneurs and Investors.
Alexandra Carina Gruber
Copyright © 2009 WILEY-VCH Verlag GmbH & Co. KGaA, Weinheim
ISBN: 978-3-527-32435-4

AUSTRIANOVA had set out to develop a cell product targeting pancreas carcinoma. The basic principle underlying the research performed by a team around Walter Günzburg from the University of Veterinary Medicine, Vienna, was to deliver the enzyme cytochrome P450 into the tumor in small capsules. Afterwards, patients were supposed to receive the cytostatic agent ifosfamide. The enzyme would then cause the chemotherapeutic agent to develop into its biologically active form right within the tumor. Positive results provided, the launch of the cell product had been planned for 2008.

AVIR Green Hills Biotechnology AG

On 9 July 2008, AVIR Green Hills Biotechnology announced the start of a clinical Phase I study with its intranasal vaccine deltaFLU (brand name of FLUVACC) against pandemic influenza. This novel generation of a live-attenuated vaccine based on the deletion of the NS1 gene has shown excellent preclinical data. In taking this step, the Viennese biotech company set yet another milestone in the development of their vaccines.

DeltaFLU lacks the pathogenicity factor NS1. Due to this deletion, the vaccine is likely to show a superior safety profile. "We are confident that the pandemic deltaFLU vaccine will be tolerated even at high doses," said Thomas Muster, founder and CEO/CSO of AVIR Green Hills Biotechnology.

Preclinical data had shown an excellent immune response even against distantly related pandemic H5 strains (cross-protection). Moreover, the vaccine triggers a long-lasting immune response – an important prerequisite for an effective vaccine.

After completion of Phase I clinical trials, AVIR Green Hills Biotechnology will start clinical Phase II studies with the pandemic and seasonal deltaFLU vaccines in 2009. Subsequently, the company plans to perform a Phase III challenge study against seasonal influenza to provide early proof of efficacy. These studies will provide the basis for Phase III field trials leading up to the application for market authorization.

Intercell AG

On 19 June 2008, Intercell announced the submission of application to register Intercell's investigational Japanese encephalitis vaccine in Canada to the Division of Biologics and Genetic Therapies Directorate of Health Canada. This submission was based on the Market Authorization Application with the European Medicines Agency EMEA and the Biological License Application with the US Food and Drug Administration FDA, submitted in December 2007.

Intercell confirmed that the regulatory approval process in the US, EU, and Australia as well as its manufacturing operations were proceeding according to plan and in line with communicated time lines for market launches in the respective territories. The vaccine for the Canadian market will be produced at Intercell's manufacturing facility in Livingston, Scotland.

On 12 May 2008, Intercell reported to have acquired Iomai Corporation to expand its late-stage product pipeline. Iomai was acquired in a fully recommended share/cash transaction for US$6.60 per share of Iomai's common stock, representing a fully diluted equity value of Iomai of approximately US$189 million (€122 million). The acquisition was to be accomplished through stock-for-stock exchange for approximately 41 percent of Iomai's current shares outstanding held by major shareholders for approximately 1.7 million Intercell shares (representing approximately four percent of Intercell's total outstanding shares) and an all-cash merger for the remaining fully diluted outstanding shares of Iomai's common stock for approximately US$119 million (€77 million).

The deal created a leading Traveler's Vaccine portfolio by combining Intercell's Japanese encephalitis vaccine with Iomai's needle-free Travelers' Diarrhea vaccine. The Travelers' Diarrhea vaccine is planned to enter pivotal Phase III trials in the first half of 2009 and is based on the only advanced needle-free vaccine patch technology in the industry. Through the transaction, Intercell gains access to yet another product-generating technology platform. It was a valuable expansion of Intercell's pipeline, leveraging the company's late-stage product development and industrialization expertise with two further programs in clinical development, including a vaccine patch for pandemic influenza in Phase II.

Intercell will finance the cash component of the transaction consideration of approximately US$119 million (€77 million) comfortably from existing reserves and expects to maintain profitability in 2008.

Intercell will also gain full rights to two additional clinical and three preclinical programs under development, the most advanced being an immunostimulant vaccine patch in Phase II for pandemic influenza. This patch is designed to enhance the immune response compared to injected pandemic influenza vaccines. If successful, it would have the effect of expanding limited vaccine supplies by allowing public health officials to use fewer or lower doses of the vaccine. The vaccine patch has recently generated positive interim immunogenicity data in a 500-subject Phase I/II study with a one-dose application. The program is funded by a grant from the United States Department of Health and Human Services.

onepharm Research & Development GmbH

On 28 May 2008, onepharm, a Vienna biotech company specialized in the development of anti-inflammatory and antiviral drugs, announced that it had acquired intellectual property from Bionetworks (Munich, Germany). Working with its collaboration partner, The Forsyth Institute (Boston, MA), onepharm has successfully confirmed, in animal models, that this technology stops bone loss in periodontitis.

The Forsyth Institute is the world's pre-eminent organization dedicated to scientific research and education in oral health and related biomedical science. Forsyth, an independent institution, leads in the discovery, communication, and application of breakthroughs in oral health and disease prevention. The Institute is affiliated with

Harvard and has collaborations with university and research organizations around the world.

"One of our new lead drug candidates acquired from Bionetworks, OPM-3023, completely inhibits bone loss in a fulminant periodontitis rat model. When we received this exiting data in spring 2008, we immediately decided to plan a first clinical trial end of this year," recalled Otto Doblhoff-Dier, COO of onepharm. "We do not know of any product on the market that shows similar effects in periodontitis." On the basis of this new data, onepharm intends to develop a new drug for the treatment of periodontal bone loss together with its research collaboration partner, The Forsyth Institute.

In parallel, onepharm is actively continuing its project work on OMP-3001, an influenza drug candidate that is being developed in cooperation with Minophagen, Japan. Some of the 25 team members are also optimizing onepharm's lead candidates OPM-3001 and OPM-3023 by chemical modifications. "We now test our modified small molecule drugs for inhibition of various specific enzymes that play key roles in inflammation," explained Oliver Szolar, CSO. "We apply computer pharmacophore models and a broad range of cell biological and biochemical methods to screen the new candidates."

References

Amis D and Stevenson H (2001) *Winning Angels – The Seven Fundamentals of Early Stage Investing*, Financial Times/Prentice Hall, Harlow.

Amram M and Kulatilaka N (1999) *Real Options – Managing Strategic Investments in an Uncertain World*, Harvard Business School Press, Boston.

Birner H (2007) *Issues and Solutions Associated with the Current Gap in Funding for Biotech Companies*, Biotech Funding Day of the Biobusiness Conference, Geneva.

Blau J (2002) Tech Slump Sinks Germany's Neuer Markt. *Spectrum IEEE*, **39**, 24–5.

Blaydon C and Wainwright F (2007) Debate Private Equity: Changing Europe for the Better? *European Business Forum*, **Spring 2007**, 22.

Bottazzi L and Da Rin M (2002) Venture capital in Europe and the Financing of Innovative Companies. *European Venture Capital*, April 2002, 231–69.

Boyle P and Boyle F (2001) *Derivatives – The Tools that Changed Finance*, Risk Books, London.

Britten N (1995) Education and Debate; Qualitative Research: Qualitative Interviews in Medical Research. *British Medical Journal*, **311**, 251–3.

Broderson H (2005) Virtual Reality: the Promise and Pitfalls of Going Virtual. *bioentrepreneur*, doi: 10.1038/bioent 881.

Burrill S (2007) *Biotech 2007 – Life Sciences: a Global Transformation*, Burrill Life Sciences Media Group, San Francisco.

Burrough B and Helyar J (2003) *Barbarians at the Gate: the Fall of RJR Nabisco*, Collins, New York.

Capurro R (2000) Hermeneutics and the Phenomenon of Information, in *Metaphysics, Epistemology, and Technology. Research in Philosophy and Technology*, **19**, 79–85.

Dewasthaly S (2007) *Vaccines – Linking Vaccine Development with Market Opportunities. Challenges and Opportunities in the Vaccine Market*, Jacob Fleming Group, London.

Dollinger MJ (2003) *Entrepreneurship – Strategies and Resources*, Pearson Education, New Jersey.

Drucker P (1970) Entrepreneurship in the Business Enterprise. *Journal of Business Policy*, **1**, 3–12.

Ernst & Young (2007) *Beyond Borders. The Global Biotechnology Report*, Ernst & Young, London.

European Venture Capital Association (2007a) *Guide on Private Equity and Venture Capital for Entrepreneurs (an EVCA special paper published online)*, EVCA, Brussels, http://www.evca.eu.

European Venture Capital Association (2007b) *Yearbook 2007*, EVCA, Brussels.

Ferruolo S, Sandercock C and Charapp D (2002) Let's make a Deal: In-licensing Agreements for Biotech Demand Due Diligence From All Parties. *Legal Times*, **XXV**.

Frei P (2006) *Assessment and Valuation of High Growth Companies*, Haupt, Bern.

Frei P (2007) *Valuation Principles*. Biovalley Investors' Lunch, http://www. venturevaluation.ch.

Biotech Funding Trends: Insights from Entrepreneurs and Investors.
Alexandra Carina Gruber
Copyright © 2009 WILEY-VCH Verlag GmbH & Co. KGaA, Weinheim
ISBN: 978-3-527-32435-4

Frei P and Leleux B (2004) Valuation – What You Need to Know. *Nature Biotechnology*, **22**, 1049–51.

Frommann H (2007) Debate Private Equity: Changing Europe for the Better? *European Business Forum*, **Spring 2007**, 18–9.

Froschauer U and Lueger M (2003) *Das qualitative Interview. Zur Praxis interpretativer Analyse sozialer Systeme*, UTB, Vienna.

Gaspar J-M (2007) Debate Private Equity: Changing Europe for the Better? *European Business Forum*, **Spring 2007**, 16–7.

Gillinger R (2007) Intercell – die aktuelle Bewertung ist ambitioniert. *Wirtschaftsblatt*, **5**, 15.

Giudici G and Roosenboom P (2004) *The Rise and Fall of Europe's New Stock Markets*, Lavoisier, Cachan.

Gund P (1977) Three-dimensional pharmacophoric pattern searching. *Progress in Molecular and Subcellular Biology*, **5**, 117–143.

Hellmann T (2006) *Financing the Entrepreneurial Venture*, Entrepreneurial Finance Module of the Master of Science Program, Danube University Krems, Austria.

Herzog S (2007) *Frühlingserwachen: Europäisches Venture Capital steht vor einer Renaissance (published online)*, Förderland, Wissen für Gründer und Unternehmer, http://www.foerderland.de.

Hodgson J (2006) *Biotechnology in Europe. Comparative Study for EuropaBio*, EuropaBio, http://www.europabio.org.

International Financial Services London (2006) *Private Equity*, http://www.ifsl.org.uk.

International Financial Services London (2007) *Private Equity*, http://www.ifsl.org.uk.

Kiernan M, Kiernan N and Goldberg J (2003) *Using Standard Phrases in Qualitative Interviews (Tipsheet 69)*, PA Penn State Cooperative Extension, University Park.

Knight K (1967) A Descriptive Model of the Intra-Firm Innovation Process. *Journal of Business of the University of Chicago*, **40**, 478–96.

Koller T, Goedhart M and Wessels D (2005) *McKinsey & Company: Measuring and Managing the Value of Companies*, John Wiley & Sons, Hoboken.

Küpper HA (2006) *Biotechnology: 25 Years*, Global LifeScience Ventures, Munich.

Lang J and the Cambridge Entrepreneurship Centre (2001) *The High-Tech Entrepreneur's Handbook. How to Start and Run a High-tech Company*, Financial Times/Prentice Hall, London.

Lerner J and Gompers P (2001) A Note on Private Equity Partnership Agreements. *Harvard Business School. Note 9-294-084.*

Lerner J, Hardymon F and Leamon A (2005) *Venture Capital and Private Equity*, John Wiley & Sons, Hoboken.

Levine DS (2005) Exit Strategies put Biotech Firms Between Rock and a Hard Place. *San Francisco Business Times*. October 28, 2005.

Lindinger A (2005) *Finanzierung österreichischer Biotech-Unternehmen unter besonderer Berücksichtigung von Venture Capital*, Master Thesis, Fachhochschule Kufstein, Tirol, Austria.

Marchart J (2007) *Aufbruchsstimmung im österreichischen Beteiligungskapitalmarkt*, Annual Meeting of the Austrian Private Equity and Venture Capital Organization, Vienna.

Mays N and Pope C (1995a) Qualitative Research: Rigour and Qualitative Research. *British Medical Journal*, **311**, 109–12.

Mays N and Pope C (1995b) Qualitative Research: Observational Methods in Health Care Setting. *British Medical Journal*, **311**, 182–4.

Metrick A (2007) *Venture Capital and the Finance of Innovation*, John Wiley & Sons, Hoboken.

Pappas MG (2002) *The Biotech Entrepreneur's Glossary*, Pappas, Shrewsbury.

Patton MQ (1987) Depth Interviewing, in *How to Use Qualitative Methods in Evaluation*, Sage, Newbury Park, 108–43.

Pisano GP (1997) *The Development Factory: Unlocking the Potential of Process Innovation*, Harvard Business School Press, Boston.

Pisano GP (2006) So zahlt sich Biotechnologie endlich aus. *Harvard Business Manager*, **11**, 68–85.

Ritter JR (1998) Initial Public Offerings. *Contemporary Finance Digest*, **2**, 5–30.

Robbins-Roth C (2001) *From Alchemy to IPO: the Business of Biotechnology*, Perseus, Cambridge.

Romaine K (2007) Venture Performance: Up, Up and Away. *Unquote*, **89**, 16.

Scrip (2007a) *Scrip World Pharmaceutical News*, Informa UK, London, p. 11:3278.

Scrip (2007b) *Scrip World Pharmaceutical News*, Informa UK, London, p. 12:3279.

Scrip (2007c) *Scrip World Pharmaceutical News*, Informa UK, London, p. 18:3281.

Scrip (2008). *Scrip World Pharmaceutical News*, Informa UK, London, p. 6:3332.

Sewell M (1998) *The Use of Qualitative Interviews in Evaluation*, University of Arizona, http://ag.arizona.edu/fcs/ cyfernet/cyfar/Intervu5.htm.

Smith RC (2007) Debate Private Equity: Changing Europe for the Better? *European Business Forum*, **Spring 2007**, 15–6.

Thomson Pharma (2007) *The Ones to Watch*, © Thomson Reuters, London, http://www.thomsonpharma.com.

Zwettler M (2007) Österreichs Biotech-Szene: Konsolidierung? Aufbruch! *Chemiereport*, **6**, 18–22.

Appendix A: List of Abbreviations

AB	Antibody
AG	Aktiengesellschaft (*English:* Corporation)
AIM	Alternative Investment Market
AIP	Antigen Identification Program
API	Active Pharmaceutical Ingredient
ARD	American Research and Development
ATCC	American Type Culture Collection
AVCO	Austrian Private Equity and Venture Capital Organization
AWS	Austria Wirtschaftsservice Gesellschaft mbH (*English:* Promotional Bank of the Republic of Austria)
BIO	Biotechnology Industry Organization
BOKU	Universität für Bodenkultur Wien (*English:* University of Natural Resources and Applied Life Sciences, Vienna)
CC	Corporate Communications
CEO	Chief Executive Officer
CEPRES	Center of Private Equity Research
CFO	Chief Financial Officer
COO	Chief Operating Officer
CP	Convertible Preferred Stock
CRO	Contract Research Organization
CSE	Copenhagen Stock Exchange
CSL	Commonwealth Serum Laboratories
CSO	Chief Scientific Officer
CVC	Corporate Venture Capital
CYP	Cytochrome P450
DCF	Discounted Cash Flow
EBIT	Earnings before Interests and Taxes
EBITDA	Earnings before Interests, Taxes, Depreciation, Amortization
EEC	European Economic Community
EMEA	European Medicines Agency
ERA	European Research Area

Biotech Funding Trends: Insights from Entrepreneurs and Investors.
Alexandra Carina Gruber
Copyright © 2009 WILEY-VCH Verlag GmbH & Co. KGaA, Weinheim
ISBN: 978-3-527-32435-4

EU	European Union
EVCA	European Venture Capital and Private Equity Association
FCF	Free Cash Flow
FDA	Food and Drug Administration
'three Fs'	Family, Friends, and Fools
FFF	Forschungsförderungsfond (*English:* Austrian Research Promotion Fund), predecessor of the FFG
FFG	Forschungsförderungsgesellschaft Austria (Austrian Research Promotion Agency)
FOP	Fibrodystrophia Ossificans Progressiva
FP7	Seventh Framework Program for Research and Technical Development of the EU
FSE	Frankfurt Stock Exchange
GDP	Gross Domestic Product
GmbH	Gesellschaft mit beschränkter Haftung (*English:* Limited Corporation)
GMP	Good Manufacturing Practice
GP	General Partner
GSK	GlaxoSmithKline
HIV	Human Immunodeficiency Virus
HR	Human Resources
HYBLIB	Human Monoclonal Antibodies from a Library of Hybridomas
IFSL	International Financial Services London
IMBA	Institute of Molecular Biotechnology of the Austrian Academy of Sciences GmbH
IMP	Research Institute of Molecular Pathology (Vienna)
IP	Intellectual Property
IPO	Initial Public Offering
IT	Information Technology
IUPAC	International Union of Pure and Applied Chemistry
JE	Japanese Encephalitis
JRC	Joint Research Centre for Non-nuclear Research
KgaA	Kommanditgesellschaft auf Aktien (*English:* Limited Commercial Partnership)
KKR	Kohlberg Kravis Roberts
LBO	Leveraged Buy-out
LP	Limited Partner
LSE	London Stock Exchange
M&A	Merger and Acquisition
MBI	Management Buy-in
MBO	Management Buy-out
MERV	Melanoma-associated Endogenous Retrovirus
n	Number
NCE	New Chemical Entity
NIH	National Institutes of Health
NPO	Non-profit Organization

NPV	Net Present Value
PCR	Polymerase Chain Reaction
PD	Pharmacodynamics
PE	Private Equity
P/E ratio	Price/Earnings Ratio
PK	Pharmacokinetics
QA	Quality Assurance
RP	Redeemable Preferred Stock
R&D	Research and Development
R&D&I	Research, Development, and Innovation
SARS	Severe Acute Respiratory Syndrome
SME	Small and Medium-sized Enterprises
SSE	Stockholm Stock Exchange
SWX	Swiss Stock Exchange
TFP	Technology Financing Program of the AWS
'three Fs'	Family, Friends, and Fools
TU	Technische Universität Wien (*English:* Vienna University of Technology)
UBC	University of British Columbia, Vancouver
UCB	Union Chimique Belge
UCLA	University of California, Los Angeles
UK	United Kingdom
US	United States
USP	Unique Selling Proposition
VC	Venture Capital
WHO	World Health Organization
WWFF	Wiener Wirtschaftsförderungsfond (*English:* Vienna Business Agency)
ZIT	Zentrum für Innovation und Technologie (*English:* The ZIT Center for Innovation and Technology, Vienna)

Appendix B: Glossary

A

Acquisition: When one company takes over another and clearly establishes itself as the new owner, the purchase – usually with stock, cash, or a combination of the two – is called an acquisition. Legally, the target company ceases to exist, the buyer 'swallows' the business, and the buyer's stock continues to be traded.

Adjusted present value (APV): A variant of the net present value (NPV) approach that is particularly applied in cases when a company's level of indebtedness changes or it has past operating losses that can be used to offset tax obligations.

After-market performance: The price level of a newly issued stock after its initial public offering (IPO). After-market performance begins on the first day of trading on the exchange. Typically, the after-market performance is measured through the lock-up period, which usually ranges between three and nine months after the IPO date. This allows for the market to 'digest' the insider shares that might be sold quickly after the lock-up period expires. To the company management and its employees, the after-market performance is vital. If the company can reach and sustain a higher market valuation than originally estimated by the underwriting syndicate in open market trading, equity fundings will be better affordable than other methods of raising capital.

Alternative Investment Market (AIM): A stock market regulated by the London Stock Exchange for young, high-growth companies.

Angel investor (US)/Business angel (UK): A wealthy individual who invests in young, high-growth entrepreneurial companies. Although angels can sometimes perform functions identical those of venture capitalists (e.g., through advice, contacts, or hands-on work), they usually invest their own capital rather than that of institutional investors.

Antibody: An immunoglobulin, or specialized immune protein, produced by special white blood cells to bind and neutralize a foreign substance, or antigen, introduced

Biotech Funding Trends: Insights from Entrepreneurs and Investors.
Alexandra Carina Gruber
Copyright © 2009 WILEY-VCH Verlag GmbH & Co. KGaA, Weinheim
ISBN: 978-3-527-32435-4

into the body. Bacteria, viruses, and other invaders stimulate the production of antibodies.

Anti-dilution provision: The right of current shareholders to maintain their fraction of ownership in a company by buying a proportional number of shares of any future issue of common stock. It therefore protects an investor from dilution resulting from later issues of stock at a lower price than the investor originally paid.

Antigen: A substance that enters the body and stimulates the production of an antibody to fight what the immune system perceives as an invader.

Asset: An item listed on the left side of a balance sheet that has been acquired by the company in an objectively measurable transaction and has specific future economic benefit, such as additional purchasing power, cash, or the ability to generate revenues.

Atypical silent partnership: This model is based on the participation of an equity donor on the yearly company profits or losses as well as on the value increase of this company. The repayment of the equity investment usually takes place after 5–7 years at the end of the legal duration period. On the one hand, the model allows a tax-advantage to those individuals who invest part of their own profits in high-risk areas, such as the biotech industry. On the other hand, this financing form provides the entrepreneur with early and rather easy access to capital without any interference from the investors' side.

B

Balance sheet: A financial statement that specifies the financial condition of a business as of a given point in time. It includes (current/non-current) assets and liabilities as well as stockholders' equity and represents a statement of the basic accounting equation.

Bankruptcy: The state of a person or company unable to repay debts. If the bankrupt entity is a company, the ownership of its assets passes from the stockholders to the bondholders. Shareholders are always the last in line to get paid if the company goes bankrupt. Secure creditors get first grabs at the proceeds from liquidation.

Basic research: The study and research on pure science that is meant to increase our scientific knowledge base. This type of research is often purely theoretical with the intent of increasing our understanding of certain phenomena or behavior but does not seek to solve or treat these problems.

Biotechnology: The manipulation of micro-organic plant or animal cells to produce materials or products which can be used in daily life.

Board of Directors: A group of individuals, elected annually by the stockholders of a corporation to represent the interests of those stockholders.

Bond: A security issued by a borrower and obligating the issuer to make specified payments to the holder over a specific time period. A coupon bond obligates the issuer to make interest, or coupon payment, over the life of the bond and to repay the face value at maturity.

Book value: The value of an account, a company, or a share of stock as indicated by the balance sheet.

Break-even point (BEP): The point at which cost or expenses and income are equal so that there is no net loss or gain.

Bridge financing: A method of financing, used to maintain liquidity while waiting for an anticipated and reasonably expected inflow of cash.

Burn rate: The rate at which a new company uses up its venture capital to finance overhead before generating positive cash flow from operations. In other words, it is a measure of negative cash flow.

Business model: The combination of factors that describe the business, including the market the business will serve, the perceived value delivered to the customer which determines profitability per unit of sale, and the sustaining factors which will allow the company to drive long-term growth.

Business plan: A document that describes the key components of the business which is used internally and as a capital-raising tool. It usually includes an executive summary, a product or service description, a market and competitive overview, production/operations, the management team, the financial data and projections, and the financial structure of the company.

C

Capital: The term refers to funds (generally cash) generated by a company to support its operations. In a very general sense, the term describes funds produced through issuing either debt or equity securities, whereas the capital section of the balance sheet normally refers to owners' or stockholders' equity.

Capital structure: The term refers to the mixture of equity and debt that a firm has raised.

Cash flow: The movement of cash associated with a company's operating, investing, and financing activities. Cash flow involves inflows and outflows of cash. A company's cash flow is considered strong if it can generate large amounts of cash relatively quickly. Cash flow is an important part of solvency.

CEPRES: Abbreviation for the Center of Private Equity Research. It was founded in 2001 as a cooperation between VCM Capital Management GmbH and the chair of banking and finance at the Johann Wolfgang Goethe-University of Frankfurt am Main. Its purpose is to gain insight into the future trends of the private equity industry via quantitative analyses of private equity market data.

Challenge study: In vaccine development, a challenge study is a double-blind, placebo-controlled study of the safety and protective efficacy of a vaccine challenged against a live virus. The challenge study takes place in a specialized clinical trial unit. Its primary purpose is to determine whether antiviral vaccines can prevent development of symptoms in human volunteers. Each volunteer is given either the vaccine or placebo and is then exposed to, or 'challenged with,' a live virus. The volunteers will be monitored for at least one month. After immunogenicity of the vaccine has been demonstrated, a safety board will decide if the volunteers will be challenged with the live virus. The infected volunteers are housed in detached buildings and are physically isolated from both staff and co-volunteers.

Co-investment: Either the syndication of a private equity financing round or an investment by an individual general or limited partner alongside a private equity fund in a financing round.

Common stock: The equity typically held by management and founders. Typically, at the time of an initial public offering, all equity is converted into common stock. These equity claims are the last to be paid upon any liquidation of the company.

Competitive advantage: Anything that allows a company to charge prices above marginal costs.

Convertible debt/bond: A type of bond that can be converted into shares of stock in the issuing company, usually at some pre-announced ratio. It is a hybrid security with debt- and equity-like features. Although it typically has a low coupon rate, the holder is compensated with the ability to convert the bond to common stock, usually at a substantial premium to the stock's market value.

Convertible preferred (CP) stock: Equity that can either be turned in for its redemption value or converted to common stock.

Cooperation: Alliance type in form of a cooperation in a specific area with shared risk between two parties.

Corporate venture capital (CVC): Venture capital (VC) investments by corporations. An initiative by a corporation to invest either in young firms outside the corporation or in business concepts originating within the corporation. These are often organized as corporate subsidiaries, not as limited partnerships. Although traditional VC seeks to maximize financial returns, corporate venture capital often mixes financial and strategic goals.

Corporation: A legal entity, separate and distinct from its stockholders, who annually elect the Board of Directors, which represents the stockholders' interests in the management of the business. The liability of the stockholders in a corporation is limited to the dollar value of their investments.

Cost of capital: A general term for the risk-adjusted discount rate of an investment project. If a company has available cash, cost of capital is the expected return foregone by investing the cash in a project rather than in any comparable financial security. If a company does not have available cash, the term refers to the cost of acquiring the cash – i.e., the cost of raising debt (effective interest) capital or the cost of raising equity capital (dilution).

Cytochrome P 450: Cytochrome P450 (abbreviated CYP, P450, infrequently CYP450) is a large and diverse superfamily of hemoproteins found in all domains of life. Cytochromes P450 use a wealth of exogenous and endogenous compounds as substrates in enzymatic reactions.

D

Deal flow: A term referring to the quantity and quality of any investment opportunity available to an investor.

Deal terms: The legal terms upon which an investment is organized (e.g., common shares with pre-emptive right).

Debt/equity ratio: The ratio of total liabilities to stockholders' equity. It indicates the extent to which a company can sustain losses without jeopardizing the interests of its creditors. Creditors possess priority claims over stockholders, and in case of liquidation, the creditors have first right to a company's assets. From an individual creditor's perspective, therefore, the amount of equity in the company's capital structure can be regarded as a buffer that helps to guarantee that there are sufficient assets available to cover individual claims.

Debt/loan capital: Capital that is invested in a business in exchange for the legal requirement to pay a certain amount as interest each year on the loan capital provided, and likely also to pay back some of the capital according to a defined schedule. In general, debt will be secured against assets.

Derivative: A financial instrument whose characteristics and value depend upon the characteristics and value of an underlier, typically a commodity, bond, equity, or currency. Examples of such derivatives include futures and options. Advanced investors sometimes purchase or sell derivatives to manage the risk associated with the underlying security, to protect against fluctuations in value, or to profit from periods of inactivity or decline.

Dilution: A reduction in the equity position of a shareholder associated with a new financing round.

Discounted cash flow (DCF) analysis: A method that determines the value of an asset as the sum of the discounted value of all cash flows projected by that asset. DCF analysis is a form of absolute valuation.

Discount rate: A term which is applied to describe the rate used in present value computations. To compute the present value of a future cash flow, for example, the cash flows are discounted at the discount rate. In this sense, the discount rate reflects the company's cost of capital – the cost of its debt and/or equity.

Diversification: An investment strategy focusing on making many investments at the same time in order to narrow the variability of the individual returns. Investors in the stock market will typically invest in a lot of different companies and industries whose returns, in the ideal case are not correlated so that any one company failure or industry downturn will be mitigated by their other investments. In early-stage investing, the variability of returns is extreme, since the likelihood of failure of the investment is high.

Divestments: Represent the realization or exiting of a private equity investment. This is generally done by selling the company, writing off the investment, or floating the company on a stock market.

Dividend: A cash payment made to the owner of a stock.

Down round: A round of financing where investors purchase stock from a company at a lower valuation than the valuation placed upon the company by earlier investors. Down rounds cause dilution of ownership for existing investors. This often means the company's founders stock or options are worth much less or even nothing at all. Unfortunately, sometimes the only option is going out of business. In this case, down rounds are necessary and welcomed.

Due diligence: The detailed review of a business plan as well as a thorough assessment of a management team prior to a private equity investment. This process includes reviewing factors such as the management team, market, competition, track record, and finances.

E

Early-stage financing: In the preseed or seed phase, financing mostly starts with the 'three Fs,' i.e., family, friends, and fools. Afterwards, it is usually the government that steps in with grants and loans, followed by business angels, and early-stage venture capitalists.

Earnings: Also called net income or net profits; it is a term referring to the difference between revenue and expenses.

Earnings before interest and taxes (EBIT): A measurement of a firm's profitability before adjusting for interest expenses or tax obligations. This measure is very often applied to compare individual companies with different levels of indebtedness.

Earnings before interest, taxes, depreciation, amortization (EBITDA): The acronym for earnings plus interest expense, taxes, depreciation, and amortization.

Earnings per share: The best known of all financial ratios is derived by dividing net income through the average number of common shares outstanding.

Endowment funds: Organizations chartered to invest money for specific purposes.

Entrepreneur: The owner or founder of an unquoted business. It often relates to an enterprising person setting up or growing a new company which requires capital.

Entrepreneurship: The assumption of risk and responsibility in designing and implementing a business strategy or starting a business.

Equity: The term refers to the ownership stake in a company. Equity holders in a company own common stocks that have been issued by that company. Two rights are associated with owning a common stock, i.e., the right to vote for the Board of Directors at the annual shareholders' meeting and the right to receive dividends if they are declared by the board.

European Medicines Agency (EMEA): A European agency for the evaluation and approval of medicinal products in the member states of the European Union.

European Venture Capital Association (EVCA): A member-based, non-profit trade association established in 1983 and based in Brussels. The EVCA represents, promotes, and protects the interests of the European private equity and venture capital industry.

Exit: The term has two related but distinct meanings. First, an exit refers to the sale or initial public offering (IPO) of a company. Second, an exit refers to the sale of an investor's stake in a company. Both definitions are not always the same (e.g., a venture capitalist must usually hold his company's stock for at least six months after an IPO).

F

Financial engineering: A cross-disciplinary field which relies on mathematical finance, numerical methods, and computer simulation to make trading, hedging, and investment decisions as well as facilitating the risk management of those decisions. By using derivatives, financial engineering allows to manage risk and create customized financial instruments.

Financing round: This term refers to a provision of capital by a private equity/venture capital group to a firm. Since venture capital organizations usually provide capital in stages, a typical venture-backed firm will receive multiple financing rounds over many years.

First round (Series A): The first round of investments by venture capitalists.

Flavivirus: Flavivirus is a genus of the family Flaviviridae, which includes the West Nile virus, dengue virus, tick-borne encephalitis virus, yellow fever virus, hepatitis C and several other viruses which may cause encephalitis. Flaviviruses are characterized by a common size, symmetry, nucleic acid, and appearance in the electron microscope. They are transmitted through the bite of an infected mosquito or tick.

Float: The equity share placed on the market in the event of an initial public offering (IPO).

Food and Drug Administration (FDA): A unit of the US government that is responsible for the regulation of prescription drugs.

Free cash flow (FCF): A measure of financial performance, calculated as operating cash flow minus capital expenditures. In other words, free cash flow represents the cash that a company is able to generate after laying out the money required to maintain or expand its asset base.

Fund: This term refers to a pool of capital which is raised periodically by a private equity/venture capital organization. It is usually set up in the form of limited partnerships, and has a life span of ten years, though extensions of several years are often possible.

Fund of funds: A fund investing in other private equity/venture capital funds rather than in operating companies. These funds are often set up by an investment bank or investment adviser.

G

Gearing: A measure of financial leverage, demonstrating the degree to which a firm's activities are funded by owner's funds versus creditor's funds. As for most ratios, an acceptable level is determined by its comparison to ratios of companies in the same industry. The best known examples of gearing ratios include the debt-to-equity ratio (total debt/total equity), times interest earned (EBIT/total interest), equity ratio (equity/ assets), and debt ratio (total debt/total assets). A company with a high gearing, i.e., more long-term liabilities than shareholder equity, is considered speculative.

General partner (GP): The term refers to an investment manager of a limited-partnership venture capital fund.

Going concern: An entity that is expected to exist into the foreseeable future. No financial problems indicating financial failure over the planning horizon are apparent.

Good manufacturing practice (GMP): A term that is recognized worldwide for the control and management of manufacturing and quality control testing of foods and pharmaceutical products.

Government bonds: A bond issued by a national government and denominated in the country's own currency. They are usually referred to as risk-free bonds and are given the highest rating available.

Government funds: A term used to describe venture capital funds organized by government bodies, or else programs to make venture-like financings with public funds.

Grants: Funds given to tax-exempt non-profit organizations or local governments by foundations, corporations, governments, small business, and individuals. Most grants are made to fund a specific project and require some level of reporting.

Gross domestic product (GDP): This term describes the monetary value of all finished goods and services that are produced within a country's borders in a specific time period, usually on an annual basis. It includes all private and public consumption, government outlays, investments, and exports less imports that occur within a defined territory.

H

H5N1: Influenza A virus subtype H5N1, also known as A(H5N1) or simply H5N1, is a subtype of the influenza A virus which can cause illness in humans and many other animal species. According to the International Committee on Taxonomy of Viruses (ICTV) database (2006), its virus code is 00.046.0.01. A bird-adapted strain of H5N1, called HPAI A(H5N1) for 'highly pathogenic avian influenza virus of type A of subtype H5N1,' is the causative agent of H5N1 flu, commonly known as 'avian influenza' or 'bird flu.' It is enzootic in many bird populations, especially in Southeast Asia.

I

Incubator: A company or facility designed to foster entrepreneurship and help start-up companies, usually technology-related, to grow through the use of shared resources, management expertise, and intellectual property.

Initial public offering (IPO): A company's first offer to sell stock to the public. An investment bank typically underwrites these offerings. Historically, IPOs have been the most lucrative exits for venture capitalists.

Intellectual property (IP): A term used to describe intangible assets, such as knowledge, brand names, or patents.

Internal rate of return (IRR): A term which is equivalent to the compound rate of return of an investment. The discount rate that makes the net present value (NPV) equal to zero is the IRR.

L

Late-stage financing: In the later stages of development, strategic alliances with corporate companies are key to success, as are deals with late-stage venture capitalists or banks. Last but not least important is the exit, usually in form of an initial public offering (IPO), trade sale, or management buy-out (MBO).

Lead compound: In drug discovery, the term refers to a compound that has pharmacological or biological activity and whose chemical structure is used as a starting point for chemical modifications in order to improve its potency, selectivity, or pharmacokinetic parameters. Lead compounds are often found in high-through-put screenings or are secondary metabolites from natural sources.

Lead investor: An investor who is either the first, largest or most active investor in an early-stage financing round. The lead investor can sometimes, though unintention-ally, influence investment decisions of other investors.

Leveraged buy-out (LBO): This term describes an acquisition of a firm or business unit, typically in a mature industry, with a significant amount of debt. The debt is then repaid according to a clearly defined schedule that binds most of the firm's cash flow.

Leverage effect: A general term describing a financial ratio that compares some form of owner's equity (or capital) to borrowed funds. The higher a company's degree of leverage, the more the company is considered risky.

Leverage ratio: This term describes the ratio of debt to total capitalization of a firm.

Liability: The probable future sacrifice of economic benefits that derives from present obligations of a particular entity to transfer assets or provide services to other entities in the future as a result of past events or transactions.

License (=R&D licensing agreement): A contract between a technology provider (licensor) and a user of that technology (licensee). The license agreement (either an in-licensing or an out-licensing deal) may be exclusive or non-exclusive.

Licensee: In a licensing agreement, the party who receives the right to use a specific technology or product in exchange for payments.

Licensor: In a licensing agreement, the party who receives payments in exchange for providing the right to use a specific technology or a certain product that it owns.

Limited partner (LP): In a private equity fund, the limited partners provide the capital, which is then invested by the general partner.

Liquidation: A term describing the process of selling assets for cash. When compa-nies go through liquidation, they normally sell their existing assets for cash which is used to pay off creditors in order of priority. Any remaining cash is distributed to the shareholders.

Liquidation preference: In its simplest meaning, the order in which different security holders receive their payments in the event of a liquidation. In a more complex meaning, some preferred stock holders may dispose over an excess liquidation preference, which guarantees a multiple of the aggregate purchase price in case of a liquidation.

Liquid market: A term describing a market with a high degree of liquidity, often resulting from a large number of buyers and sellers. Changes in supply or demand have a smaller impact on price. A liquid market is the opposite of a thin market.

M

Management buy-in (MBI): This situation occurs when a manager or a management team from outside the company raises the necessary finances, buys the company, and becomes the company's new management. A management buy-in team often competes with other purchasers in the search for a suitable business.

Management buy-out (MBO): A term refering to a leveraged buy-out (LBO) which is initiated by an existing management team, which then solicits the involvement of a private equity group.

Market capitalization (market cap): This term refers to the equity market value. It is calculated by multiplying the number of shares outstanding by the current market price of one share.

Merger: Happens when two firms, often of about the same size, agree to go forward as a single new company rather than remain separately owned and operated (also referred to as a 'merger of equals').

Mergers and Acquisitions (M&A): The phrase mergers and acquisitions refers to the aspect of corporate strategy, corporate finance, and management dealing with the buying, selling, and combining of different companies that can aid, finance, or help a growing company in a given industry grow rapidly without having to create another business entity.

Mezzanine: This term either refers to a private equity financing round shortly before an initial public offering (IPO) or an investment that uses subordinated debt with fewer privileges than bank debt but more than equity and often has attached warrants.

Milestone payments: In a licensing agreement, the payments which are either made by the licensee to the licensor at specified times in the future or when certain technological/business objectives have been accomplished. In drug development, typical milestones are the achievement of Phase II, Phase III, and FDA/EMEA approval.

Monte Carlo analysis (Monte Carlo simulation): Term describing the estimation of the expected value by computing the average from a simulated sample of a large number of observations.

Multiples: This term is equal to comparables analysis. It refers to the valuation of an asset by using data on similar assets (multiples analysis, method of multiples).

N

Net present value (NPV): A valuation method that calculates the expected value of one or more future cash flows and discounts them at a rate that presents the cost of capital (which will vary with the riskiness of the cash flow).

Networking: In venture capital investing, networking is the process of meeting people in order to create deal flow or identify co-investors.

Net worth: For a company, the term refers to total assets minus total liabilities. Net worth is an important determinant of the value of a company, considering it is composed primarily of all the money that has been invested since its inception, as well as the retained earnings for the duration of its operation. It is also called shareholders' equity or net assets.

New chemical entity (NCE): A chemical molecule developed by an innovator company in the early drug discovery stage, which after undergoing clinical trials could translate into a drug that could be a cure for some disease. Synthesis of an NCE is the first step in the process of development of a drug. Once the synthesis of the NCE has been completed, companies can either go for clinical trials on their own or license the NCE to another company.

Non-profit organizations (NPO): A legally constituted organization whose primary objective is to support or to actively engage in activities of public or private interest without any commercial or monetary profit purposes. NPOs are active in a wide range of areas, including the environment, humanitarian aid, animal protection, education, the arts, social issues, charities, early childhood education, health care, politics, religion, research, sports, or other endeavors.

NS1: The NS1 influenza protein is produced by the internal protein-encoding, linear negative-sense, single-stranded RNA, NS gene segment detected in Influenzavirus A, Influenzavirus B, and Influenzavirus C. The NS1 protein of the highly pathogenic avian H5N1 viruses circulating in poultry and waterfowl in Southeast Asia is believed to be responsible for the enhanced virulence of the strain.

O

Oligopoly: A market dominated by a small number of participants who are able to collectively exert control over supply and market prices. This refers to a market condition in which sellers are so few that the actions of any one of them will materially affect price and have a measurable impact on competitors.

Options/Stock options: Financial instruments which give the holder the right to buy an underlying instrument (e.g., common stock) at an agreed amount.

Orphan drug: A term referring to a product that treats rare diseases which affect fewer than 200 000 Americans. The Orphan Drug Act was signed into law in the US in 1983. The intent of the Orphan Drug Act is to encourage the research, development, and approval of products that treat rare diseases. Because medical research and development of drugs to treat such diseases is financially disadvantageous, companies that do so are rewarded with tax reductions and marketing exclusivity (a 'monopoly') on that drug for an extended time (seven years post-approval). A similar status exists in the European Union, administered by the

Committee on Orphan Medicinal Products of the European Medicines Agency (EMEA).

Outsourcing: This term refers to contracts for certain projects with a fixed timeline and limited risk for the partner in order to cut costs (e.g., the procuring of a service, such as the manufacturing of a vaccine by an outside supplier).

P

Partnership: (1) Generally speaking, a term that refers to a relationship of two or more entities conducting business for mutual benefit. (2) A legal, binding contract defining the association of two or more persons in a business or professional relationship.

Patent: This term describes a government grant of rights to one or more discoveries for a set period, based on a specified set of criteria.

Personalized medicine: The use of an individual's genetic information (e.g., by doctors) to customize the prevention, detection, or treatment of a disease.

Pharmacodynamics (PD): The study of the biochemical and physiological effects of drugs on the body or on microorganisms or parasites within or on the body, the mechanisms of drug action, and the relationship between drug concentration and effect. Pharmacodynamics is often summarized as the study of what a drug does to the body.

Pharmacokinetics (PK): A branch of pharmacology dealing with the reactions between drugs or synthetic food ingredients and living structures (e.g., tissues, organs) through studies on the absorption, distribution, metabolism, and elimination of the compound in a living system. In short, pharmacokinetics is the study of what the body does to the drug.

Pharmacophore models: The term describes a set of structural features in a molecule which is recognized at a receptor site and is responsible for that molecule's biological activity (Gund, 1977). The IUPAC definition of a pharmacophore is 'an ensemble of steric and electronic features that is necessary to ensure the optimal supramolecular interactions with a specific biological target and to trigger (or block) its biological response.' In modern computational chemistry, pharmacophores support the process of identifying the essential features of one or more molecules with the same biological activity. Databases of diverse chemical compounds can be searched for molecules which share the same features.

Phase I: The first phase of clinical trials for a drug after preclinical testing usually in animals. Phase I trials test for the safety of the drug in healthy volunteers. Phase I clinical testing also seeks to determine how the pharmaceutical is absorbed, distributed, how long it is active in the body, how it is metabolized by the body, and how and when it is excreted (pharmacokinetics).

Phase II: The second phase of clinical trials for a drug. Phase II trials test for efficacy and safety of the drug in patients who suffer from the relevant disorder. In addition, side effects, effective dose strength, and dosing schedule are determined.

Phase III: The third and final phase of human trials for a drug before approval status. Phase III trials study efficacy and safety of the drug using patients who have the relevant disorder. They differ from Phase II trials as these clinical studies are usually much larger, take longer, and are much more expensive.

Platform technologies: Technologies (e.g., in preclinical development, genomics, bioimaging, structural biology) that can be used to facilitate a wide range of application-based activities. The use of appropriate platform technologies can reduce costs and avoid unnecessary duplication of facilities, increase international R&D competitiveness, and provide an environment of effective networking and collaboration.

Portfolio company: A company that has received one or more venture capital investments, but has not yet been exited.

Post-money valuation: The value or total price of a start-up including the new investment. Thus, pre-money value plus investment equals post-money value.

Preclinical: The stage of drug testing that precedes clinical trials.

Preferred stock: Shares with preferential rights over common stock with respect to any dividends or payments in association with the liquidation of the company. Moreover, the preference shares may also have additional rights, such as the ability to block mergers and/or to displace the management team.

Pre-money valuation: The value or total price of a company start-up before the investment is included.

Present value: This term refers to a technique which is used to place a value, as of the present day, on a set of future cash flows. It is computed by discounting future cash flows at an interest rate that reflects a company's cost of capital.

Price/Earnings (P/E) ratio: The ratio of the firm's share price to the firm's earnings per share (calculated by dividing net income through shares outstanding). It is also referred to as the P/E multiple.

Private companies: These companies have equity shares that are not listed and traded on the public stock exchange.

Private equity: In its broadest meaning, private equity includes all investments that cannot be resold in public markets. In a more narrow term, private equity refers to a class of investments, managed by private equity firms, which make investments in venture capital, leveraged buy-outs, mezzanine, or distress.

Proceeds: The amount of value (in cash and in stock) that is received in an exit.

Q

Quality assurance (QA): Generally speaking this term refers to the activity of providing evidence required to establish quality in work and to ensure that activities which require good quality are being performed effectively. Hence, the term describes those planned or systematic actions necessary to provide enough confidence that a product or service will satisfy the given requirements for quality. Particularly biotechnology companies face some unique quality and validation challenges during the production of biopharmaceutical products.

R

Random walk: Describes the notion that stock price changes are random and unpredictable.

Real options: This term describes an option on a real asset such as plant, land, or machinery. In contrast, financial options are options on financial assets.

Recapitalization: A company incurring significant additional debt by repurchasing stocks through a buyback program or by distributing a large amount of dividend among its current shareholders.

Red biotechnology: An emerging field that refers to the use of organisms for the improvement of medical processes. It includes the design of organisms to manufacture pharmaceutical products, such as antibiotics or vaccines, the engineering of genetic cures through genomic manipulation, and its use in forensics through DNA profiling.

Redeemable preferred (RP) stock: Preferred stock that pays a redemption value on liquidation but does not offer the holder the option of conversion to common stock. The return to the investor derives from the face value of the share which is paid out at the specified time when the company has to redeem the shares.

Redemption value: The amount paid to the holder of preferred stock on redemption.

Royalties: Payments made by one company (the licensee) to another company (the licensor) in exchange for the right to use IP or other assets owned by the licensor.

S

Sarbanes–Oxley: The Sarbanes–Oxley Act of 2002, also known as the Public Company Accounting Reform and Investor Protection Act of 2002, is a US federal law enacted on 30 July 2002 in response to a number of major corporate and accounting scandals, including those affecting Enron, Tyco International, or WorldCom. The legislation establishes new or enhanced standards for all US public company boards, management, and public accounting firms.

Secondary offering (follow-on offering): This term refers to an issuance of stock subsequent to the company's initial public offering. A follow-on offering can be dilutive, non-dilutive, or a mixture of both. A secondary offering is an offering of securities by a shareholder of the company (as opposed to the company itself, which is a primary offering).

Second round (Series B): The second occurrence of a venture capital investment.

Seed capital: Capital that will finance a company at the conceptual stage.

Serial entrepreneur: An entrepreneur who starts a new business after having started and exited a previous business venture. Serial entrepreneurs often treat entrepreneurship as a profession and have a high level of expertise in creating new ventures, but not necessarily in operating these ventures in the long term.

Serial investor: An angel or venture capital investor who participates in many early-stage investment deals.

Smart money: Money from investors who make profitable investment moves at the right time, no matter what the investing environment; more generally, money from investors who are highly experienced or have inside information.

Soft money: 'One-time' funding from governments or organizations for a project or special purpose.

Spin-off: A new organization or entity formed by a split from a larger one or as a new company formed from a university research group. In a traditional spin-off, it means the creation of an independent company through the sale or distribution of new shares of an existing business or division of a parent company. A spin-off is a type of divestiture.

Start-up: The activity of an early-stage company, typically with two or more of the following characteristics: a management team, equity capital, a prototype, or a sale.

Stock market flotation: This term describes going public to raise equity financing and to allow the original owners and early investors to realize some of their gains.

Strategic alliance: A formal relationship formed between two or more parties to pursue a set of agreed goals or to meet a critical business need while remaining independent organizations. Partners may provide the strategic alliance with resources such as products, distribution channels, manufacturing capability, project funding, capital equipment, knowledge, expertise, or intellectual property. The alliance is a cooperation or collaboration which aims for a synergy where each partner hopes that the benefits from the alliance will be greater than those from individual efforts. The alliance often involves technology transfer (access to knowledge and expertise), economic specialization, shared intellectual property, shared expenses, and shared risk.

Subordinated debt: Also known as a subordinated loan, subordinated bond, subordinated debenture, or junior debt, this term describes a debt which ranks after other debts should a company fall into receivership or be closed.

Syndicate: In some cases, private equity funds will come together in order to form a 'financial syndicate' for an investment. This may happen in case the risks are high or if the amount of capital required in the operation is substantial. One of the investment funds will represent the group in the syndicate's dealings with the entrepreneur.

Syndication: The joint purchase of shares by two or more private equity organizations or the joint underwriting of an offering by two or more investment banks.

T

Term sheet: Gives a preliminary outline of the structure of a private equity/venture capital partnership or stock purchase agreement, which is usually agreed to by the key parties before the formal contractual language is negotiated.

Trade sale: A European term for exiting an investment by a private equity/venture capital group by selling it to a corporation. It normally entails the disposal of a company's shares or assets. This may refer to a strategic buyer who intends to grow his business or to a financial buyer who wants to generate a financial return on his invested capital at the time of exit.

Turnarounds: A favorable reversal in the fortunes of a company, a market, or the economy at large. Hence, turnarounds are investments into a distressed company or a company where value can be unlocked as a result of a one-time opportunity, such as changing industry trends or government regulations. Stock market investors speculating that a poorly performing company is about to show a marked improvement in earnings might profit from such a turnaround.

U

Unique selling proposition (USP): This term describes the characteristics and advantages that differentiate a company from its competitors and that are emphasized in marketing.

V

Valuation: The total determined value of a company, which usually derives from the price paid by new investors. The total valuation is essentially the price one would be willing to pay to buy the whole company.

Value proposition: This term refers to the fundamental proposition for customers of the early-stage company and to what they will get for their money. If customers accept to pay more for the product than it costs the company to make it, the value proposition is real.

Venture capital (VC): This term describes independently managed, dedicated pools of capital that focus on equity or equity-linked investments in privately held, high growth companies.

Venture capitalist: A general partner or associate at a private equity organization who invests capital in high-growth potential companies on behalf of a fund.

Vero cells: Viral vaccine production has traditionally depended on growing the virus either in host animals (e.g., mice, embryonated eggs), in primary cell culture, or in diploid cell lines. All of these methods carry the risk of contamination, inconsistency, long production cycles, and high production costs. Therefore, more and more new viral vaccines are produced from continuous cell lines. Vero cells are a well characterized continuous cell line derived from the African Green Monkey kidney cells, deposited at and distributed by the American Type Culture Collection (ATCC). The safety of Vero cells for the production of biologicals has been demonstrated through some 20 years of administering more than 100 million doses of vaccine worldwide in more than 60 countries. The Vero cell line technology is an ideal platform for the production of a variety of different viral vaccines, such as influenza, Ross River, and SARS corona virus.

Vesting: A provision that is included in employment agreements for restricting employees from exercising all or some of their stock options immediately. The agreement usually provides a schedule determining which percentage of shares the employee is allowed to exercise over time, known as vesting schedule.

W

Warrant: An option to buy shares of stock issued directly by a company.

White biotechnology: An emerging field within modern biotechnology that serves industry. It uses living cells, such as moulds, yeasts, or bacteria, as well as enzymes to produce goods or services.

World Health Organization (WHO): A specialized agency of the United Nations (UN) that acts as a coordinating authority on international public health. Established on 7 April 1948 and headquartered in Geneva, Switzerland, the agency inherited the mandate and resources of its predecessor, the Health Organization, which had been an agency of the League of Nations.

Appendix C

C.1 Overview of Biotech Companies Mentioned in This Book

Actelion Ltd (Switzerland)	www.actelion.com
Amgen, Inc. (US)	www.amgen.com
Antares Pharma, Inc. (US)	www.antarespharma.com
AstraZeneca PLC (UK)	www.astrazeneca.com
AVIR Green Hills Biotechnology AG (Austria)	www.greenhillsbiotech.com
BASF SE (Germany)	www.basf.com
Basilea Pharmaceutica AG (Switzerland)	www.basilea.com
Baxter International, Inc. (US)	www.baxter.com
Bayer Schering Pharma AG (Germany)	www.bayer.com
Biogen Idec (US)	www.biogenidec.com
Biological E. Ltd (India)	www.biologicale.com
Biovitrum AB (publ.) (Sweden)	www.biovitrum.com
BioXell S.p.A. (Italy)	www.bioxell.com
CARBOGEN AMCIS AG (Switzerland)	www.carbogen-amcis.com
Cellectis SA (France)	www.cellectis.com
CIBA (Switzerland)	www.ciba.com
CSL Ltd (Australia)	www.csl.com.au
Cytos Biotechnology AG (Switzerland)	www.cytos.com
DSM Fine Chemicals Austria GmbH & Co KG (Austria)	www.dsm.com/de_DE/html/dfca/dfca_home.htm
Eli Lilly, Inc. (US)	www.lilly.com
EUCODIS GmbH (Austria)	www.eucodis.com
Genaera, Inc. (US)	www.genaera.com
Genentech, Inc. (US)	www.gene.com
GenVec, Inc. (US)	www.genvec.com
Gilead Sciences, Inc. (US)	www.gilead.com
GlaxoSmithKline GSK Ltd (UK)	www.gsk.com
Henkel KGaA (Germany)	www.henkel.com
IMBA GmbH (Austria)	www.imba.oeaw.ac.at

Biotech Funding Trends: Insights from Entrepreneurs and Investors.
Alexandra Carina Gruber
Copyright © 2009 WILEY-VCH Verlag GmbH & Co. KGaA, Weinheim
ISBN: 978-3-527-32435-4

Intercell AG (Austria)	www.intercell.com
Iomai Corporation (US)	www.iomai.com
Jerini AG (Germany)	www.jerini.com
Kirin Pharma Company, Ltd (Japan)	www.kirinpharma.co.jp
Lohmann Animal Health GmbH & Co KG (Germany)	www.lah.de
Marinomed Biotechnology GmbH (Austria)	www.marinomed.com
Medical Discoveries, Inc. (US)	www.medicaldiscoveries.com
MedImmune, Inc. (US)	www.medimmune.com
Merck & Co., Inc. (US)	www.merck.com
Merck KGaA (Germany)	www.merck.de
Merck Serono International SA (Switzerland)	www.merckserono.net
Minophagen Pharmaceutical Co., Ltd (Japan)	www.minophagen.co.jp
MorphoSys AG (Germany)	www.morphosys.com
Nabriva Therapeutics AG (Austria)	www.nabriva.com
NextPharma Technologies (UK)	www.nextpharma.com
Novartis International AG (Switzerland)	www.novartis.com
Nycomed International Management GmbH (Switzerland)	www.nycomed.com
Onepharm Research and Development GmbH (Austria)	www.onepharm.com
Oramed Pharmaceuticals Inc. (Israel)	www.oramed.com
Pelias Biomedizinische Entwicklungs AG (Austria)	www.pelias.com
Pfizer, Inc. (US)	www.pfizer.com
Pliva d.d. – a member of the Barr Group (Croatia)	www.pliva.com
Sandoz International GmbH (Germany)	www.sandoz.com
Sanochemia Pharmazeutika AG (Austria)	www.sanochemia.at
Sanofi–Aventis SA (France)	www.sanofi-aventis.com
Sanofi Pasteur SA (France)	www.sanofipasteur.com
Siegfried Ltd (Switzerland)	www.siegfried.ch
Speedel Ltd (Switzerland)	www.speedel.com
Statens Serum Institute SSI (Denmark)	www.ssi.dk
Union Chimique Belge SA (Belgium)	www.ucb-group.com
VTU Engineering GmbH (Austria)	www.vtu.com
Wyeth Pharmaceuticals, Inc. (US)	www.wyeth.com

C.2 Overview of Venture Capital Companies Mentioned in This Book

Acceres Beteiligungsmanagement GmbH & Co KG (Deutschland)	www.acceres-vc.de
Apax Partners Ltd (UK)	www.apax.com
Athena Wien Beteiligungen AG (Austria)	www.ipo-austria.at
Gamma Capital Partners Beratungs- und Beteiligungs AG (Austria)	www.gamma-capital.com
Global Life Science Ventures (Germany)	www.glsv-vc.com
Kapital & Wert Vermögensverwaltung AG (Austria)	www.kapital-wert.com
Life Science Partners (Netherlands)	www.lspv.com
MPM Capital (US)	www.mpmcapital.com
Nomura Holdings, Inc. (Japan)	www.nomura.com
Novartis Venture Funds AG (Switzerland)	www.venturefund.novartis.com
Pontis Venture Partners Management GmbH (Austria)	www.pontisventure.at
Ryan Group Holdings (Ireland)	http://ryangroupholdings.com
SR–One Ltd (GlaxoSmithKline)	www.srone.com
Techno Venture Management TVM Capital GmbH (Germany)	www.tvm-capital.com
Temasek Holdings (Private) Ltd (Singapore)	www.temasekholdings.com.sg

Appendix D: Questionnaires

D.1 Questionnaire for the Entrepreneur

I. Introduction

- **If you search for an investor, what are your most important criteria and do you search for VC firms investing in early stage or which cover the whole development phases?**

    Most important criteria: _____

    ☐ early stage only   ☐ whole development

  Would you prefer a national, regional or international fund?

    ☐ national   ☐ regional   ☐ global

    Why? _____

- **What should be the major capabilities & strengths of an entrepreneur/venture capitalist today?**

    Major capabilities VC: _____

    _____

    Major capabilities entrepreneur: _____

    _____

- **How do you try to solve the difficulty related to the dual role of a scientist being entrepreneur / general manager at the same time and can you give an example?**

    Possible solution: _____

    _____

    Example: _____

    _____

Biotech Funding Trends: Insights from Entrepreneurs and Investors.
Alexandra Carina Gruber
Copyright © 2009 WILEY-VCH Verlag GmbH & Co. KGaA, Weinheim
ISBN: 978-3-527-32435-4

II. Financing models

- **What stage in biotech are you currently in?**

  ☐ seed   ☐ early   ☐ late stage

- **What are – from your personal perspective – the most important financing models for European Biotech companies today?**

  ☐ A) Venture capital   ☐ B) Corporate venture capital   ☐ C) Private equity

  ☐ D) Strategic alliances (Pharma)   ☐ E) Mezzanine capital   ☐ F) Debt financing (banks)

  ☐ G) Government grants   ☐ H) EU grants   ☐ I) Non profit organizations / foundations

  ☐ J) Angel investors   ☐ K) Others: _____

Is there any particular new and creative approach?

  New, creative approach: _____

What are the major strengths / weaknesses of your top three financing approaches?

  Ranking: ☐ 1 (most important)   ☐ 2   ☐ 3 (least important)

What is your personal ranking for the top three approaches?

  Rank 1: Strength _____

  　　　　Weakness _____

  Rank 2: Strength _____

  　　　　Weakness _____

  Rank 3: Strength _____

  　　　　Weakness _____

Did you already receive venture capital?

  ☐ Yes   ☐ No

- **With regards to debt financing:**

How do you value the role of banks as capital investors?

  Role: _____

What are the hurdles?

  Hurdles: _____

III A. Strategic alliances / Early stage

- **How important are strategic alliances between universities / their scientists and VCs in Europe and can you give an example from your own experience?**

  ☐ High importance   ☐ Medium   ☐ Low

  Example: _____

What can scientists and VCs do to establish/enhance such networks?

  Suggestions: _____

- **With regards to the different available government funds (e.g. Austria, Germany, Switzerland): how do you value the different government supports in financing biotech (grants, loans, equity participations)?**

  Grants: _____

  Loan: _____

  Equity participation: _____

What is the most important one? _____

- **Do you feel that funds are used in the most efficient way? Is there any area for improvement?**

  ☐ Yes _____

  ☐ No _____

  Area for improvement _____

- **Do you have any partnership/cooperation with angel investors?**

  ☐ Yes _____

  ☐ No _____

What is the role of angel investing in Europe?

  ☐ High importance   ☐ Medium   ☐ Low

III B. Strategic alliances / Late stage

- **What should a pharmaceutical company offer to an entrepreneur to establish / enhance strategic alliance networks?**

    Ideas:_____

    _____

- **What is your preferred exit strategy?**

    ☐ IPO   ☐ MBO   ☐ Acquisition   ☐ Other: _____

- **Would markets like AIM be an interesting alternative?**

    ☐ Yes _____

    please specify: _____

    ☐ No _____

    please specify: _____

- **What is from your own perspective the most prominent recent success story for a European Biotech company that went public?**

    Best practice example: _____

- **What do you think is the best approach to structure the board of directors?**

    _____

    _____

IV. Biotech

• **What are – from your own perspective – the most promising biotech research areas today and why?**

☐ Hemato/Oncology   ☐ Cardiovascular/Metabolic disease   ☐ Neurology

☐ Respiratory   ☐ Immunology   ☐ Pain treatment

☐ Medical technology/device _____

☐ Diagnostics _____

☐ Others: _____

Why? _____

• **What roles do vaccines (therapeutic, preventive use) play in the global biotech & VC area today and why?**

☐ High importance   ☐ Medium   ☐ Low

Why? _____

Do you think this role changed over the last 5-10 years?

☐ Yes   ☐ No

• **Are the business models from "classic biotech" totally applicable to vaccines development?**

☐ Applicable   ☐ Non-applicable

Why? _____

• **Do you see any similarities or differences in the financing models between classic biotech and vaccines?**

Major Similarities: _____

Major Differences: _____

V. Conclusion

• **How essential is it for the investor to receive milestone payments during the investment period ?**

   ☐ High importance   ☐ Medium   ☐ Low

• **What is from your own perspective the primary cause why many biotech companies fail?**

   Primary cause: ―――――――――――――――――――――――――――――

• **Which country in Europe is a best practice example in funding biotech start ups?**

   Best practice country: ――――――――――――――――――――――――――――

• **What are the major differences in VC financing in Europe versus US from your personal perspective?**

   Major differences: ―――――――――――――――――――――――――――――

   ――――――――――――――――――――――――――――――――――――――

D.2 Questionnaire for the Venture Capitalist

I. Introduction

- Shall the venture capitalist be acting in a pure national, regional or global partnership/network and why?

  ☐ national   ☐ regional   ☐ global

  Why? _____

Is there any difference to the past?

  ☐ yes   ☐ no

- What should be the major capabilities & strengths of a venture capitalist / an entrepreneur today?

  Major capabilities VC: _____

  _____

  Major capabilities entrepreneur: _____

  _____

- How do you try to solve the difficulty related to the dual role of a scientist being entrepreneur / general manager at the same time and can you give an example?

  Possible solution: _____

  _____

  Example: _____

  _____

II. Financing models

• **What stages in biotech are you interested in?**

  ☐ seed   ☐ early   ☐ late stage

How advanced must the Biotech company be so that you decide to step in?

  ☐ seed   ☐ early   ☐ late stage

What are your most important entry criteria?

  Most important entry criteria: _____

• **What are – from your personal perspective – the most important financing models for European Biotech companies today?**

  ☐ A) Venture capital   ☐ B) Corporate venture capital   ☐ C) Private equity

  ☐ D) Strategic alliances (Pharma)   ☐ E) Mezzanine capital   ☐ F) Debt financing (banks)

  ☐ G) Government grants   ☐ H) EU grants   ☐ I) Non profit organizations / foundations

  ☐ J) Angel investors   ☐ K) Others: _____

Is there any particular new and creative approach?

  New, creative approach: _____

What is your personal ranking for the top three approaches?

  Ranking: ☐ 1 (most important)   ☐ 2   ☐ 3 (least important)

What are the major strengths / weaknesses of your top three financing approaches?

  Rank 1: Strength _____

  　　　　Weakness _____

  Rank 2: Strength _____

  　　　　Weakness _____

  Rank 3: Strength _____

  　　　　Weakness _____

How many biotech companies – according to our own experience – do receive venture capital?

  Approximate value (%): _____

• **Are there any financing particularities / differences for early versus late stage development?**

      Early financing models: _____

      Late financing models: _____

      Difference: _____

• **What are the general valuation techniques that you are using for early/late stage biotech companies?**

      _____

• **With regards to equity financing: What is your preferred deal structure?**

      Preferred deal structure: _____

What is your preference – majority/minority investment and why?

      ☐ Majority   ☐ Minority

      Why? _____

Has your preference changed over time?

      ☐ Yes   ☐ No

• **With regards to debt financing: How do you value the role of banks as capital investors?**

      Role: _____

What are the hurdles?

      Hurdles: _____

III A. Strategic alliances / Early stage

- **How important are strategic alliances between VCs and universities / their scientists in Europe and can you give an example from your own experience?**

    ☐ High importance   ☐ Medium   ☐ Low

    Example: _____

  What can VCs and scientists do to establish/enhance such networks?

    Suggestions: _____

- **With regards to the different available government funds (e.g. Austria, Germany, Switzerland): how do you value the different government supports in financing biotech (grants, loans, equity participations)?**

    Grants: _____

    Loan: _____

    Equity participation: _____

  What is the most important one? _____

- **Do you feel that funds are used in the most efficient way? Is there any area for improvement?**

    ☐ Yes _____

    ☐ No _____

    Area for improvement _____

- **Do you have any partnership/cooperation with angel investors?**

    ☐ Yes _____

    ☐ No _____

  What is the role of angel investing in Europe?

    ☐ High importance   ☐ Medium   ☐ Low

III B. Strategic alliances / Late stage

• **How do you value partnerships between VCs and corporate ventures in Europe and can you give an example from your own experience?**

      ☐ High importance   ☐ Medium   ☐ Low

      Example: _____

What can VCs / corporate venture funds / entrepreneurs do to establish / enhance such networks?

      Suggestion: _____

• **What are your preferred exit strategies in biotech and can you give an example?**

      ☐ IPO   ☐ MBO   ☐ Acquisition   ☐ Other: _____

      Example: _____

Would markets like AIM be an interesting alternative?

      ☐ Yes

      please specify: _____

      ☐ No

      please specify: _____

• **What is your most recent success story for a European Biotech company which went public?**

      Best practice example: _____

• **What do you think is the best approach to structure the board of directors?**

      _____

      _____

IV. Biotech

• **What are – from your own perspective – the most promising biotech research areas today and why?**

☐ Hemato/Oncology   ☐ Cardiovascular/Metabolic disease   ☐ Neurology

☐ Respiratory   ☐ Immunology   ☐ Pain treatment

☐ Medical technology/device _____

☐ Diagnostics _____

☐ Others: _____

Why? _____

• **What roles do vaccines (therapeutic, preventive use) play in the global VC area today and why?**

☐ High importance   ☐ Medium   ☐ Low

Why? _____

Has this role changed over the last 5-10 years?

☐ Yes   ☐ No

Can you give an example for a successful vaccine development company that you were supporting with VC?

Example: _____

• **Are the business models from "classic biotech" totally applicable to vaccines development?**

☐ Applicable   ☐ Non-applicable

Why? _____

• **Do you see any similarities or differences in the financing models between classic biotech and vaccines?**

Major Similarities: _____

Major Differences: _____

V. Conclusion

- **How essential is it for the investor to receive revenues during the investment period?**

  ☐ High importance   ☐ Medium   ☐ Low

And how often does this happen?

  Frequency: ☐ High   ☐ Medium   ☐ Low

If so, would you also reinvest this money?

  ☐ Yes   ☐ No

  Why?_____

- **What is from your own perspective the primary cause why many biotech companies fail?**

  Primary cause: _____

- **Which country in Europe is a best practice example in funding biotech start ups?**

  Best practice country: _____

- **What are the major differences in VC financing in Europe versus US from your personal perspective?**

  Major differences: _____

  _____

D.3 Questionnaire for the Group of Other Investors

I. Introduction

- Shall the investor be acting in a pure national, regional or global partnership/network and why?

  ☐ national   ☐ regional   ☐ global

  Why? _____

Is there any difference to the past?

  ☐ yes   ☐ no

- What should be the major capabilities & strengths of a venture capitalist / an entrepreneur today?

  Major capabilities VC: _____

  _____

  Major capabilities entrepreneur: _____

  _____

- How do you try to solve the difficulty related to the dual role of a scientist being entrepreneur / general manager at the same time and can you give an example?

  Possible solution: _____

  _____

  Example: _____

  _____

II. Financing models

- **What stages in biotech are you interested in?**

    ☐ seed   ☐ early   ☐ late stage

How advanced must the Biotech company be so that you decide to step in?

    ☐ seed   ☐ early   ☐ late stage

What are your most important entry criteria?

    Most important entry criteria: _____

- **What are – from your personal perspective - the most important financing models for European Biotech companies today?**

    ☐ A) Venture capital   ☐ B) Corporate venture capital   ☐ C) Private equity

    ☐ D) Strategic alliances (Pharma)   ☐ E) Mezzanine capital   ☐ F) Debt financing (banks)

    ☐ G) Government grants   ☐ H) EU grants   ☐ I) Non profit organizations / foundations

    ☐ J) Angel investors   ☐ K) Others: _____

Is there any particular new and creative approach?

    New, creative approach: _____

What are the major strengths / weaknesses of your top three financing approaches?

    Ranking: ☐ 1 (most important)   ☐ 2   ☐ 3 (least important)

What is your personal ranking for the top three approaches?

    Rank 1: Strength _____

          Weakness _____

    Rank 2: Strength _____

          Weakness _____

    Rank 3: Strength _____

          Weakness _____

How many biotech companies – according to our own experience – do receive venture capital?

    Approximate value (%): _____

• **Are there any financing particularities / differences in early versus late stage development?**

Early financing models: _____

Late financing models: _____

Difference: _____

• **What are the general valuation techniques that you are using for early/late stage biotech companies?**

_____

• **With regards to equity financing: Is there any particular new and creative approach?**

Preferred deal structure: _____

What is your preference – majority/minority investment and why?

☐ Majority   ☐ Minority

Why? _____

Has your preference changed over time?

☐ Yes   ☐ No

• **With regards to debt financing: How do you value the role of banks as capital investors?**

Role: _____

What are the hurdles?

Hurdles: _____

III A. Strategic alliances / Early stage

• **How important are strategic alliances between universities / their scientists and VCs in Europe and can you give an example from your own experience?**

    ☐ High importance   ☐ Medium   ☐ Low

    Example: _____

What can scientists and VCs do to establish/enhance such networks?

    Suggestions: _____

• **With regards to the different available government funds (e.g. Austria, Germany, Switzerland): how do you value the different government supports in financing biotech (grants, loans, equity participations)?**

    Grants: _____

    Loan: _____

    Equity participation: _____

What is the most important one? _____

• **Do you feel that funds are used in the most efficient way? Is there any area for improvement?**

    ☐ Yes _____

    ☐ No _____

    Area for improvement _____

• **Do you have any partnership/cooperation with angel investors?**

    ☐ Yes _____

    ☐ No _____

What is the role of angel investing in Europe?

    ☐ High importance   ☐ Medium   ☐ Low

III B. Strategic alliances / Late stage

- **How do you value partnerships between VCs and corporate ventures in Europe and can you give an example from your own experience?**

    ☐ High importance   ☐ Medium   ☐ Low

    Example: _____

    What can VCs / corporate venture funds / entrepreneurs do to establish / enhance such networks?

    Suggestion: _____

- **What are your preferred exit strategies in biotech and can you give an example?**

    ☐ IPO   ☐ MBO   ☐ Acquisition   ☐ Other: _____

    Example: _____

    Would markets like AIM be an interesting alternative?

    ☐ Yes

    please specify: _____

    ☐ No

    please specify: _____

- **What is your most recent success story for a European Biotech company which went public?**

    Best practice example: _____

- **What do you think is the best approach to structure the board of directors?**

    _____
    _____

IV. Biotech

• **What are – from your own perspective – the most promising biotech research areas today and why?**

    ☐ Hemato/Oncology   ☐ Cardiovascular/Metabolic disease   ☐ Neurology

    ☐ Respiratory   ☐ Immunology   ☐ Pain treatment

    ☐ Medical technology/device _____

    ☐ Diagnostics _____

    ☐ Others: _____

    Why?_____

• **What roles do vaccines (therapeutic, preventive use) play in the global biotech & VC area today and why?**

    ☐ High importance   ☐ Medium   ☐ Low

    Why? _____

  Do you think this role changed over the last 5-10 years?

    ☐ Yes   ☐ No

• **Are the business models from "classic biotech" totally applicable to vaccines development?**

    ☐ Applicable   ☐ Non-applicable

    Why _____

• **Do you see any similarities or differences in the financing models between classic biotech and vaccines?**

    Major Similarities: _____

    Major Differences: _____

V. Conclusion

- **How essential is it for the investor to receive revenues during the investment period?**

  ☐ High importance   ☐ Medium   ☐ Low

  And how often does this happen?

  Frequency: ☐ High   ☐ Medium   ☐ Low

  If so, would you also reinvest this money?

  ☐ Yes   ☐ No

  Why?_____

- **What is from your own perspective the primary cause why many biotech companies fail?**

  Primary cause: _____

- **Which country in Europe is a best practice example in funding biotech start ups?**

  Best practice country: _____

- **What are the major differences in VC financing in Europe versus US from your personal perspective?**

  Major differences:_____

  _____

Index

Biotech Funding Trends: Insights from Entrepreneurs and Investors.
Alexandra Carina Gruber
Copyright © 2009 WILEY-VCH Verlag GmbH & Co. KGaA, Weinheim
ISBN: 978-3-527-32435-4